Fatigue and Fracture of Weldments

Uwe Zerbst · Mauro Madia
Benjamin Schork · Jonas Hensel
Pawel Kucharczyk · DesireTchoffo Ngoula
Didi Tchuindjang · Julian Bernhard
Carla Beckmann

Fatigue and Fracture of Weldments

The IBESS Approach for the Determination of the Fatigue Life and Strength of Weldments by Fracture Mechanics Analysis

Springer

Uwe Zerbst
Bundesanstalt für Materialforschung
und -prüfung (BAM)
Berlin, Germany

Mauro Madia
Bundesanstalt für Materialforschung
und -prüfung (BAM)
Berlin, Germany

Benjamin Schork
Center for Engineering Materials—State
Materials Testing Institute Darmstadt
(MPA), Institute for Materials
Technology(IfW)
Technische Universität Darmstadt
Darmstadt, Hessen, Germany

Jonas Hensel
Institute of Joining and Welding IFS
Braunschweig University of Technology
Braunschweig, Niedersachsen, Germany

Pawel Kucharczyk
IWT Solutions AG
Aachen, Nordrhein-Westfalen, Germany

DesireTchoffo Ngoula
Materials Mechanics Group
Technische Universität Darmstadt
Darmstadt, Hessen, Germany

Didi Tchuindjang
Saarbrücken, Saarland, Germany

and

Hamburg University of Technology (TUHH)
Hamburg, Germany

Julian Bernhard
SAM, LBF Darmstadt
Technische Universität Darmstadt
Darmstadt, Hessen, Germany

Carla Beckmann
Fraunhofer-Institut für Werkstoffmechanik
Freiburg im Breisgau, Baden-Württemberg,
Germany

ISBN 978-3-030-04072-7 ISBN 978-3-030-04073-4 (eBook)
https://doi.org/10.1007/978-3-030-04073-4

Library of Congress Control Number: 2018960756

This Springer imprint is published by the registered company Springer Nature Switzerland AG
The registered company address is: Gewerbestrasse 11, 6330 Cham, Switzerland

Foreword

This book gives a comprehensive and thorough guide to the readers who are lost in the forest of fatigue of weld. It presents the clear fracture mechanics and material background of fatigue of weld by starting first from fatigue crack initiation, short crack and next by discussing long cracks, crack closure, crack growth and threshold, residual stress, stress concentration, stress intensity factor, J-integral, multiple cracks, weld geometries and defects, microstructural parameters including HAZ and cyclic stress–strain behavior. The authors have succeeded in treating all these mutually interacting important parameters in a unique distinct way of analysis. The elaborate works performed by the deep scientific knowledge of the authors should be admired.

This is the timely and excellent publication in the field of weld.

Fukuoka, Japan

Yukitaka Murakami
Emeritus Professor
Kyushu University

Notice

While the authors believe that the information and guidance given in this book are correct, all parties making use of it must rely on their own skill and judgement. The authors cannot assume any liability for loss or damage caused by any error or omission in the application of the IBESS method. Any and all such liabilities are disclaimed.

The authors do not give any warranty or guarantee whatsoever that the information and guidance given in this book do not infringe the rights of any third party or can be used for any particular purpose at all. Any person intending to use the same should satisfy himself as to accuracy and the suitability for the purpose for which it is intended to be used.

Contents

Contributors

J. Baumgartner Fraunhofer Institute for Structural Durability and System Reliability LBF, Darmstadt, Germany

H. Th. Beier Materials Mechanics Group, Technische Universität Darmstadt, Darmstadt, Germany

W. Fricke Hamburg University of Technology (TUHH), Hamburg, Germany

J. Hohe Fraunhofer-Institut für Werkstoffmechanik, Freiburg, Germany

M. Kaffenberger Center for Engineering Materials—State Materials Testing Institute Darmstadt (MPA), Institute for Materials Technology (IfW), Technische Universität Darmstadt, Darmstadt, Germany

S. Münstermann RWTH Aachen University, Aachen, Germany

T. Nitschke-Pagel Institute of Joining and Welding IFS, Technical University Braunschweig, Braunschweig, Germany

M. Vormwald Materials Mechanics Group, Technische Universität Darmstadt, Darmstadt, Germany

Abbreviations

ASTM	American Society for Testing and Materials
bcc	Body-centered cubic (lattice)
BM	Base metal
BS	The British Standards Institution
CDF	Crack driving force
CG	Coarse grain (HAZ)
COV	Coefficient of variation (ratio of standard deviation σ and mean or expected value μ of a statistical distribution)
CP	Cap position at a weld (Fig. 23)
CPCA	Compression pre-cracking constant amplitude
CPLR	Compression pre-cracking load reduction
C(T)	Compact tension specimen (fracture mechanics testing)
cyc	Cyclic
FAD	Failure assessment diagram
FAT	FAT class, stress range referring to 2×10^6 loading cycles
fcc	Face-centered cubic (lattice)
FCP	Fatigue crack propagation
FG	Fine grain (HAZ)
HAZ	Heat-affected zone
IIW	International Institute of Welding
ISO	International Organization for Standardization
L	Longitudinal
LC	*Long* crack
MAG	Metal active gas (welding)
NASGRO	Computer program for fatigue crack propagation, provided by NASA
OM	Strength overmatching ($\sigma_{YW} > \sigma_{YB}$)
p	Plasticity-corrected
p	Primary
PWHT	Post-weld heat treatment
R-curve	Crack resistance curve (monotonic and cyclic version)

s	Secondary
SC	*Short* crack
SE(B)	Single notch bending specimen (fracture mechanics testing)
T	Transverse
TIG	Tungsten inert gas (welding)
UM	Strength undermatching ($\sigma_{YW} < \sigma_{YB}$)
WM	Weld metal

Nomenclature

a	Crack length (crack depth for surface cracks)
a_i	Initial crack depth (for fracture mechanics analysis)
a_f	Final crack depth
a_0	El Haddad parameter, Eq. (6)
a^*	Correction term for modified El Haddad's model, Eqs. (7) and (8)
a/c	Crack aspect ratio (changes during crack propagation and coalescence)
A	Elongation at fracture (uniaxial tensile test)
$\|\|b\|\|$	Burgers vector
b,k	Parameters for describing a notch geometry (Fig. 74)
B	Specimen thickness (fracture mechanics specimen)
c	Half crack length at surface (semi-elliptical crack)
C,n	Fit parameters of the da/dN–ΔK curve in the Paris regime
C	Gradient for load reduction in standard FCP threshold testing, Fig. 6.
da/dN	Fatigue crack propagation rate
E	Modulus of elasticity (Young's modulus)
E'	$=E$ for plane stress and $E/(1-\nu^2)$ for plane strain conditions
f	Crack closure function in the NASGRO equation, Eq. (82)
f_σ	Coefficient in $\sigma_e - R_m$ correlation
$f(L_r)$	Plasticity correction function (monotonic loading)
$f(\Delta L_r)$	Plasticity correction function (cyclic loading)
F_i	Components of the boundary correction function of K-factor solutions
h	Weld reinforcement, Fig. 68
h	Stress triaxiality (ratio of hydrostatic stress to the equivalent stress; in Fig. 44 to the yield stress of the weld)
H	Width or half width of the weld strip (strength mis-match consideration)
HB	Hardness according to Brinell
HV	Hardness according to Vickers

i	Counting parameter; i = 1, 2, 3, …
J	J-integral (monotonic loading)
$J_{0.2;BL}$	Resistance against stable crack initiation (monotonic loading)
$J_{0.2}$	Resistance against stable crack initiation (alternative definition)
k	Depth of a secondary notch, e.g., an undercut
k	Slope of finite life (high cycle) fatigue S-N curve in double logarithmic scaling, Eq. (107)
K′, n′	Coefficients of the cyclic stress–strain curve (Ramberg–Osgood)
K	Stress intensity factor (K-factor)
K_c^J	Monotonic fracture resistance (formally derived from J-integral)
K_0	Scale parameter in three-parameter Weibull distribution
K_{mat}	Fracture resistance, monotonic loading (general term)
K_{max}	Maximum K-factor in a loading cycle
K_{med}	50% percentile value of fracture resistance in the Master Curve approach
K_{min}	Minimum K-factor in a loading cycle
K_{min}	Shift parameter in three-parameter Weibull distribution
K_{op}	K-factor at crack opening
K_r	K-factor due to residual stresses
K_r	Ordinate of the FAD ($=K/K_{mat}$)
K_I	K-factor for mode I crack opening (normal to the crack faces)
$\overline{K}$	Mean value of the K-factor (in the loading cycle)
ℓ	Section width along the weld toe (IBESS approach)
L	Weld width, Fig. 77
L_r	Ligament yielding parameter (monotonic loading)
m	Shape parameter in three-parameter Weibull distribution
m(z,a)	Weight function (K-factor determination)
M	Strength mis-match ratio (commonly σ_{YW}/σ_{YB})
N	Number of loading cycles
N	Number of specimens in a statistical test set (Figs. 29, 30, 41)
N_c	Number of loading cycles up to fracture
p,q	Fit parameters of the threshold and fracture regions of the da/dN–ΔK curve, Eqs. (4) and (80)
P	Probability
P_f	Failure probability
P_t	Roughness parameter used for specifying the secondary notch depth k in the present study, Annex D
R	Loading ratio ($=\sigma_{min}/\sigma_{max}$ or K_{min}/K_{max})
R_{eL}	Lower yield strength (materials showing a Lüders' plateau)
$R_{p0.2}$	0.2% proof strength (materials without Lüders' plateau)
R_m	Uniaxial tensile strength
SCF	(Elastic) Stress concentration factor (at weld toe surface)
T	Plate thickness
T_i	Components of the traction vector for determining the J-integral

T_p	Peak temperature during welding (Fig. 21)
T_0	Transition temperature in the Master Curve approach
u_i	Displacement vector components for determining the J-integral
U	Crack closure parameter ($\Delta K_{eff}/\Delta K$)
U	Energy dissipated in monotonic fracture mechanics test
U_{LC}	Crack closure factor for *long* cracks, independent of crack depth a
U_{SC}	Crack closure factor for *short* cracks, as function of crack extension Δa
V	Plasticity correction factor for secondary stresses
W	Specimen width or half width (fracture mechanics specimen)
W	Strain energy density (determination of the J-integral)
Y	Boundary correction function in K-factor solution
Z	Elastic follow-up factor
α	Weld flank angle, Figs. 66, 67
α_g	Constraint parameter in the NASGRO equation, Eqs. (83, 84, 88, 89)
Δa	Crack extension (monotonic and cyclic)
$\Delta \delta$	Cyclic crack tip opening displacement
ΔJ	Cyclic J-integral (cyclic loading)
ΔJ_e	Elastic part of the cyclic J-integral
ΔK	K-factor range (K_{max}–K_{min})
ΔK_{eff}	Effective K-factor range (= K_{max}–K_{op})
ΔK_p	(Formally) plasticity-corrected ΔK, Eq. (29)
ΔK_{th}	Fatigue crack propagation threshold
$\Delta K_{th,eff}$	Intrinsic fatigue crack propagation threshold (no crack closure effect)
$\Delta K_{th,LC}$	Fatigue crack propagation threshold in the *long* crack regime
$\Delta K_{th,op}$	Crack closure component of the fatigue propagation threshold
$\Delta K_{th,SC}$	Fatigue crack propagation threshold in the (physically) *short* crack regime
$\Delta \varepsilon$	Strain range (ε_{max}–ε_{min})
$\Delta \varepsilon_{ref}$	Cyclic counterpart to ε_{ref}
$\Delta \varepsilon_{i,j}$	Components of the strain tensor range
ΔL_r	Ligament plasticity factor (cyclic loading)
$\Delta \sigma$	Stress range (σ_{max}–σ_{min})
$\Delta \sigma_{app}$	Applied stress range (refers to cross section without crack)
$\Delta \sigma_{ref}$	Cyclic counterpart to σ_{ref}
$\Delta \sigma_{i,j}$	Components of the stress tensor range
$\Delta \sigma_e$	Endurance limit range
ΔT_i	Components of the traction vector range (cyclic J-integral determination)
ΔW	Cyclic strain energy (cyclic J-integral determination)
Δu_i	Components of the displacement vector range
ε	Strain
ε_a	Strain amplitude (= ½ $\Delta \varepsilon$)
ε_{ref}	Reference strain (reference stress method)
ε_{ij}	Components of the strain tensor

η_p	Geometry function in monotonic J-integral testing
μ	Mean or expected value of statistical distributions
ρ	Weld toe radius, Figs. 64, 65
ρ	Notch tip radius (Fig. 74)
ε_{max}	Maximum strain in the fatigue cycle
ε_{min}	Minimum strain in the fatigue cycle
$\gamma(R)$	Coefficient for mean stress correction of σ_e, Eqs. (105) and (106)
Γ	Integration path for J-integral determination, Fig. 46
ν	Poisson's ratio
σ	Stress
σ	Standard deviation of statistical distributions
σ_a	Stress amplitude ($= \frac{1}{2}\,\Delta\sigma$)
σ_b	Bending stress
σ_{bs}	Bending stress due to misalignment as referred to the membrane stress σ_m, Sect. 2.6
σ_e	Endurance limit (stress amplitude), defined for $N = 10^7$ in IBESS
σ_i	Polynomial coefficients of the through-thickness stress profile, Eqs. (35) and (36)
σ_{ij}	Components of the stress tensor
σ_f	Flow stress in the original NASGRO equation [$= \frac{1}{2}(\sigma_Y + R_m)$]
σ_m	Membrane stress (tension loading)
σ_m	First invariant of stress tensor (Fig. 44)
σ_{max}	Maximum stress in the fatigue cycle
σ_{min}	Minimum stress in the fatigue cycle
σ_n	Maximum remote stress (Eq. 31)
σ_{ref}	Reference stress (reference stress method)
σ_0	Reference yield stress, Sect. 2.2.3.2.6
$\sigma_{0,m}$	Reference yield stress (membrane; tension loading)
$\sigma_{0,b}$	Reference yield stress (bending loading)
σ_1	First principal stress (refers to mode I loading in the case of a crack)
$\sigma(z)$; $\sigma_1(z)$	Through-thickness stress profile
σ_Y	Yield strength, general (either R_{eL} or $R_{p0.2}$)
σ'_Y	(Stabilized) cyclic yield strength
σ_{YB}	Yield strength of base metal
σ_{YW}	Yield strength of weld metal
σ^p	Primary stress
σ_r	Residual stress
σ_r^T	Transverse residual stress
σ^s	Secondary stress

Abstract

The acronym IBESS stands for "Integrale Bruchmechanische Ermittlung der Schwingfestigkeit von Schweißverbindungen" which, translated from German, means "integral fracture mechanics determination of the fatigue strength of welds." The method introduced in this study is the outcome of a German research cluster in which eight partners were involved. A list of them is found at the end of this study. The IBESS method is characterized by a number of partially novel aspects and elements of fracture mechanics applied to the evaluation of fatigue strength of welds. The most important ones are:

- Determination of fatigue crack propagation for *mechanically/physically short* and *long* cracks.
- Determination of an elastic–plastic crack driving force for the treatment of *mechanically short* cracks. To that purpose, an analytical expression for the cyclic J-integral was developed and validated against finite element results.
- The gradual buildup of the crack closure phenomenon is determined by using cyclic R-curves which describe the crack size dependency of the fatigue crack propagation threshold in the *physically short* crack growth regime.
- A physically meaningful initial crack size is defined for total life consideration. It is based on a two-criterion approach. Based on a cyclic R-curve analysis, the crack size at crack arrest is determined as a lower bound. If, however, a pre-existing crack-like defect is larger than this, its dimensions define the initial crack size.
- Multiple crack propagations at the weld toe are considered.
- In conjunction with this, the variation of the weld toe geometry is considered in a stochastic model.
- As a result, both the fatigue limit (defined for 10^7 loading cycles) and the finite life (high cycle) fatigue S-N curve are obtained statistically.
- At various analysis steps, parametric equations have been developed which allow for analytical calculations instead of complete stochastic analyses based on finite elements which are unrealistic even at present.

- The method has been validated with a large number of S-N curves including two materials, three weldment types with two weld geometries, each referring to different manufacturing technologies and the as-welded and stress-relieved state.
- Although not finally solved, an extended discussion is provided on the issue of welding residual stresses including their redistribution under cyclic loading.
- A number of simplifications is proposed at lower analyses levels which, however, partly lack complete validation by now.

Introduction

The basic topic of the present study is the determination of the fatigue strength by fracture mechanics. Common fracture mechanics-based fatigue considerations, e.g., within the frame of damage tolerance considerations, are usually limited to residual lifetime determination of so-called *long* cracks which are often defined in conjunction with nondestructive testing. The extension of this concept to the total lifetime, as in the S-N curve approach, requires an adequate description of so-called *short* crack propagation which cannot be based on the ΔK concept and must consider the crack closure phenomenon as well as its gradual buildup at the *short* crack stage. The terms *long* and *short* cracks will be explained in Sect. 1.1.3. Further, it has to provide a meaningful definition of the initial crack dimensions needed in fracture mechanics and a solution for the multiple crack problems at stress levels higher than the fatigue limit as this is specific for configurations such as weldments. A fundamental discussion of these aspects beyond its application to welds is provided by Zerbst et al. (2018b, c).

The idea to apply fracture mechanics to the determination of the fatigue strength of weldments is anything but new. First attempts date to around 1970 (Maddox 1970) who wrote in 1974: "It is now widely recognized that flaws will inevitably exist in welded structures and the old idea of removing all detectable defects must be replaced by the 'fitness for purpose' design philosophy. This makes it necessary to define reliable methods of assessing the significance of flaws, particularly in the context of fatigue, …. The most promising approach to this problem lies in the use of the fracture mechanics based description of fatigue crack propagation." (Maddox 1974). At that time, a major problem was to develop appropriate K-factor solutions. Ever since effort was spent to further develop the approach. No detailed overview shall be given here; see, however, Radaj et al. (2006), Chap. 6, for a review. Note that even today, fracture mechanics besides some promising applications (see, e.g., Nykänen 2005, representative for many other studies) is still not widely used as a tool for fatigue strength determination of weldments. One reason for this is certainly that the approaches usually fail to meet the above-formulated requirements. They are based on linear elastic fracture mechanics even for short cracks and on fixed or (from S-N curves) back-calculated initial crack sizes, they usually consider one crack only, neglect the variation in the local weld geometry along the weld toe, etc.

The present study aims at a discussion of all these points and at offering possible solutions obtained within the frame of the German research cluster IBESS. It combines rather basic notes with the introduction of a methodology for the fracture mechanics-based determination of the fatigue strength of welds, but also with a critical discussion of hitherto not satisfactorily resolved problems.

Partners Involved in the IBESS Cluster Project

- Bundesanstalt für Materialforschung und -prüfung Berlin, Division 9.1 (Project coordination).
- Rheinisch-Westfälische Technische Hochschule (RWTH) Aachen, Institut für Eisenhüttenkunde (IEHK).
- Technische Universität Braunschweig, Institut für Füge- und Schweißtechnik (IFS).
- Technische Universität Darmstadt, Institut für Stahlbau und Werkstoffmechanik (IfSW), Fachgebiet Werkstoffmechanik.
- Technische Universität Darmstadt, Fachgebiet Systemzuverlässigkeit und Maschinenakustik (SzM), Fraunhofer-Institut für Betriebsfestigkeit und Systemzuverlässigkeit (LBF).
- Technische Universität, Staatliche Materialprüfungsanstalt Darmstadt und Fachgebiet und Institut für Werkstoffkunde (MPA/IfW).
- Fraunhofer-Institut für Werkstoffmechanik (IWM) Freiburg.
- Technische Universität Hamburg-Harburg, Institut für Konstruktion und Festigkeit von Schiffen (SKF).

The authors wish to thank Prof. Dr. Isabel Hadley (The Welding Institute (TWI), Great Abington, Cambridge) for her valuable discussion of the content of this book.

The authors wish also to thank the Deutsche Forschungsgemeinschaft (DFG) for the financial funding of four IBESS subprojects. Furthermore, the authors gratefully appreciate the funding of the four other subprojects by the Arbeitsgemeinschaft industrieller Forschungsvereinigungen (AiF) network, in particular

Forschungsvereinigung Stahlanwendung (FOSTA), Forschungsvereinigung Schweißen und verwandte Verfahren (DVS), Forschungsvereinigung Automobiltechnik (FAT), and Center of Maritime Technologies (CMT).

Gefördert durch:

Bundesministerium
für Wirtschaft
und Energie

aufgrund eines Beschlusses
des Deutschen Bundestages

Fatigue and Fracture of Weldments

1 Background Information

1.1 Stages of Fatigue Crack Propagation and Lifetime

1.1.1 Crack Initiation

Fatigue cracks are initiated at the smooth surface in plain specimens (i.e., at surface extrusions and intrusions caused by persistent slip bands) or—more frequently—at geometrical discontinuities (which act as geometric stress concentrators) and/or material defects such as non-metallic inclusions (which cause strain concentration zones due to their different stiffness compared to the matrix material and sometimes also chemical mis-match), see Zerbst et al. (2018b). Polak (2003) notes with respect to the crack nucleation stage: “Numerous studies have shown that in the majority of materials and under normal loading conditions, the period of crack initiation in smooth specimens without defects amounts to less than 5–20% of the fatigue life. In materials containing defects, the fraction of life spent in crack initiation is even lower. The major part of the life is spent in the growth of cracks, namely in the growth of *short* cracks.” That this statement usually is also true with respect to welds has been confirmed, e.g., by Verreman and Nie (1996) for manual fillet welds of structural steel.

Subsequent to crack initiation, fatigue crack propagation is characterized by distinct successive stages, Fig. 1, see, e.g., Tanaka (2003), Davidson et al. (2003).

1.1.2 Microstructurally Short Cracks

As long as its size is still in the order of microstructural features such as the grain size, the crack is designated as *microstructurally short*. The crack and the plastic zone ahead of its tip is completely embedded within a single or a few grains. As the

U. Zerbst et al., *Fatigue and Fracture of Weldments*,
https://doi.org/10.1007/978-3-030-04073-4_1

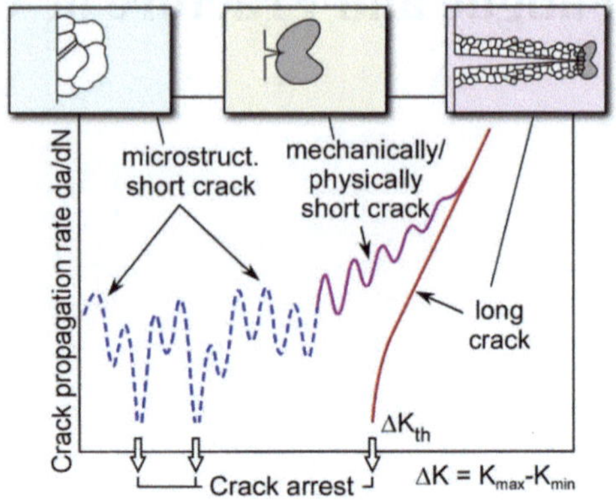

Fig. 1 Stages of fatigue crack propagation, schematic view, according to Miller (1993), slightly modified

consequence, crack growth is strongly affected by the local microstructure and characterized by (temporary) acceleration and deceleration phases or even crack arrest. Typical growth barriers are grain boundaries between adjacent grains of different crystal orientation where the generation of the plastic zone is inferred. An example is shown in Fig. 2. The microstructural crack propagation stage is terminated when the crack depth (of the initially semi-circular or semi-elliptical crack) reaches the order of one or a few grain sizes (Murakami 2002); a few, since not every grain boundary will act as a crack stopper when, e.g., the crystal orientation of two adjacent grains is not significantly different. Note that, in general terms, it is not the crack depth but the crack front length which is the critical parameter. E.g., Capelli et al. (2008), investigating single phase aluminium alloy, restrict the *microstructurally short* crack stage to a crack front length intersecting 15 grains. That also means that the above statement about the critical crack depth no longer applies for through surface cracks or long shallow surface cracks showing a small crack aspect ratio a/c. The application of classical fracture mechanics to *microstructurally short* cracks is not possible (Pippan and Riemelmoser 2003).

1.1.3 Mechanically and Physically Short Cracks

A *microstructurally short* crack which is not arrested will finally become a *mechanically short* crack. The general definition of this is a size up to the order of the plastic zone size ahead of its tip. As the consequence, the linear elastic fracture mechanics concept based on the K-factor range ΔK is not applicable but elastic-plastic concepts based on the cyclic J-integral, ΔJ, or a cyclic crack tip opening displacement, $\Delta\delta$, have to be used. This will be discussed in Sect. 2.2.

The term "*physically short* crack" refers to the observation that during that stage the crack closure phenomenon is being gradually built-up. It does not exist at the very beginning, but then the effect intensifies with increasing crack size (and crack wake

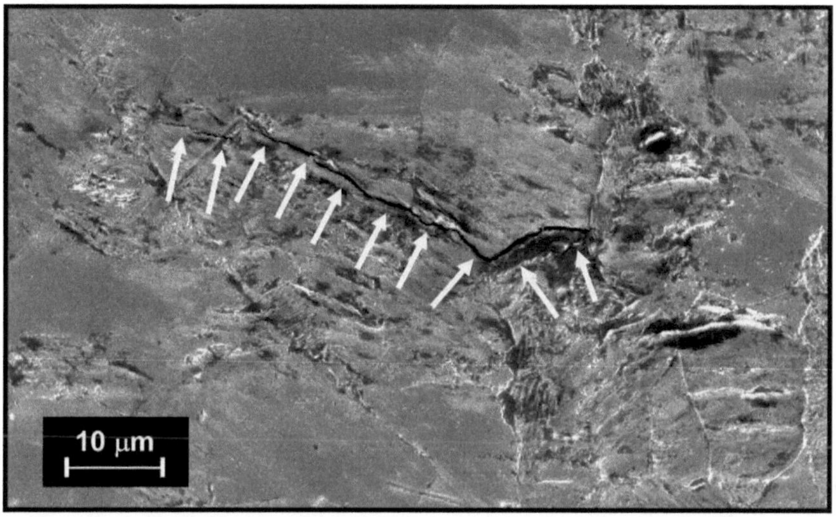

Fig. 2 Non-propagating crack in EA1N railway axle steel. The crack stopped within a grain. It experienced more than 10^8 loading cycles; according to Beretta et al. (2009)

length) and finally it reaches a stable, nearly crack size independent state when the crack becomes a *long* one. The crack closure phenomenon is the topic of Sect. 1.2.

Note that the difference in the terms "*mechanically*" and "*physically*" *short* cracks is not different crack sizes (these at least will overlap) but the different phenomena.

1.1.4 Long Cracks

A *mechanically/physically short* crack becomes a *long* one when the crack closure effect stabilizes, i.e. it becomes independent of the crack size and when the linear elastic K concept is applicable (see above). An exception is low cycle fatigue (LCF) which occurs at stress levels close to the yield strength. Excluding this exception, *long* crack propagation can be described by the da/dN–ΔK curve and the *long* fatigue crack propagation (FCP) threshold $\Delta K_{th,LC}$. The *long* crack stage is terminated by the failure of the structure, i.e., by monotonic fracture but there might also be other failure criteria associated with loss of functionality.

Figure 3 provides a scheme of the crack sizes referring to the stages discussed above. Note, however, that the numbers are strongly material dependent. Therefore, the figure should be regarded as a first impression only which roughly describes the relations for steel.

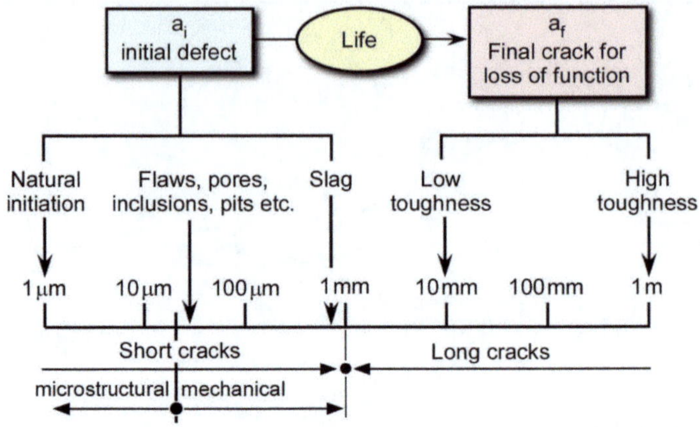

Fig. 3 Crack length scales of the life cycle of a component subjected to cyclic loading; according to Tanaka (2003)

Excursus: A Competing Definition

To avoid confusion, it has to be mentioned that a competing definition of the fatigue crack propagation stages is also in use, preferably in a practical context. A crack initiation stage is defined which is not terminated until the crack has reached a size detectable with realistic effort and a crack propagation stage beyond that crack size. For steels, the latter roughly refers to the *long* crack stage in Fig. 3.

Note that the frequent statement, that the fatigue life of welds is controlled by crack propagation rather than by crack initiation, originally referred to this nomenclature. In the past it provided the justification to apply *long* crack fracture mechanics to the fatigue life determination of welds. Although the initial crack size is reduced to a small fraction of a millimeter in more recent documents such as the IIW Guidelines (Hobbacher 2016) *long* crack fracture mechanics based on the da/dN–ΔK curve is still in use. However, this is no longer consistent in view of the information provided in Sect. 1.1.3. As in many other applications, the major part of the fatigue life of welds is used to be spent at the *short* crack stage as long as no large initial defects such as extended lack of penetration exist. Note, however, that the assessment of those defects is a task for damage tolerance rather than for safe life considerations based on S-N curve data.

1.1.5 Failure

Failure occurs when the crack reaches a critical size such that the maximum crack driving force in the loading cycle, e.g. K_{max}, reaches or exceeds the monotonic fracture resistance of the material. Besides this, other failure criteria are possible as well when the functionality of the component is impaired even if the component does not break yet.

1.2 The Crack Closure Phenomenon

1.2.1 Brief Overview

Since it can be assumed that a fatigue crack does not grow when the crack is closed, i.e., its faces are in contact during the lower part of the loading cycle at negative R ratios ($R = K_{min}/K_{max}$), only part of the stress intensity factor range ΔK ($= K_{max} - K_{min}$) will contribute to its propagation. This is designated by an effective value ΔK_{eff} ($= K_{max} - K_{op}$) with K_{op} being the K-factor at crack opening, for the definition see Fig. 4a. As a first approximation, ΔK_{eff} should be ½ ΔK for $R = -1$, however, it is observed to be even smaller, i.e., the crack is open only at stress levels above zero.[1] This is designated as "crack closure phenomenon". It is commonly described by a crack closure parameter U such that

$$\Delta K_{eff} = U \cdot \Delta K \tag{1}$$

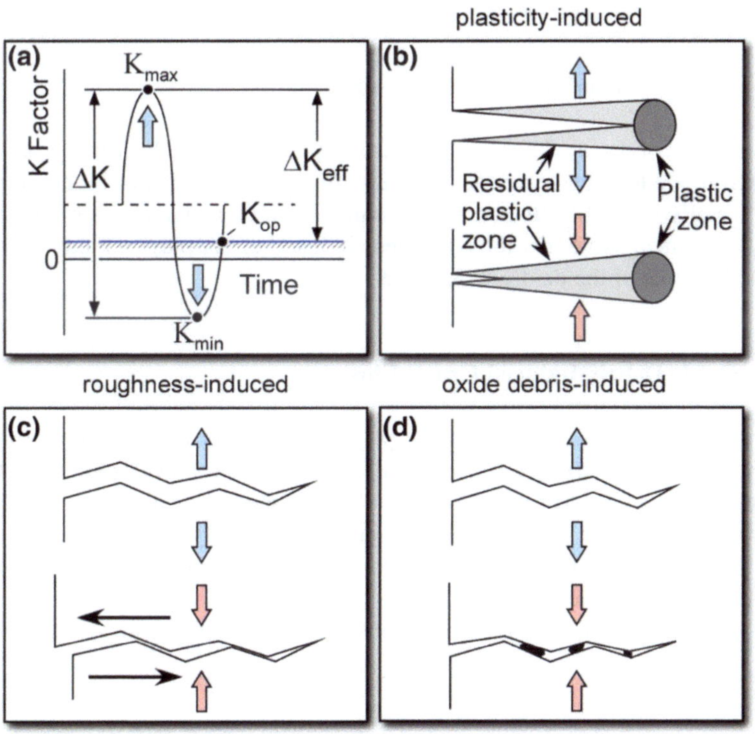

Fig. 4 Mechanisms of the crack closure phenomenon; **a** nomenclature; **b** plasticity-induced mechanism; **c** roughness-induced mechanism; **d** oxide debris-induced mechanism

[1] Sometimes also below, depending on the R-ratio and the applied stress ranges

Several mechanisms are responsible for the crack closure phenomenon. The three most important ones are the plasticity-induced, the roughness-induced and the oxide debris-induced crack closure mechanisms (Fig. 4b–d). For these and further ones see, e.g., Tanaka and Akiniwa (2003) and Suresh (2003).

1.2.2 Plasticity-Induced Crack Closure

Under load, a plastic zone is formed ahead of the crack tip. When the crack propagates, this will remain as plastically stretched material at the surfaces of the wake of the crack (Fig. 4b). The consequence is geometrical mis-match particularly at the specimen sides where the plastic zone dimension is largest due to prevailing plane stress conditions. Since the plastic zone size increases quadratically with the applied load and decreases quadratically with the yield strength of the material (for small scale yielding conditions) the plasticity-induced mechanism becomes stronger for higher maximum loads in the loading cycle and for smaller yield strengths. This is at least the case for *long* cracks.

1.2.3 Roughness-Induced Crack Closure

The roughness of the crack faces causes asperity contact (Fig. 4c) which is enhanced by mixed mode loading components (shear) due to the kinking or branching of the crack. Usually it will be affected by the grain size in that a larger grain size tends to a rougher surface. Note that the roughness-induced effect plays a major role in the threshold regime of the da/dN–ΔK curve.

1.2.4 Oxide Debris-Induced Crack Closure

This mechanism is active for corrosion-prone materials or material conditions. The crack wake surfaces are covered by oxide layers which thicken due to fretting of the crack faces at low R ratios. The locally furbished surfaces are again exposed to corrosion (Fig. 4d). This way a debris layer is gradually formed the thickness of which is in the order of the crack tip opening displacement in the threshold regime of the da/dN–ΔK curve.

All crack closure mechanisms become stronger for low and disappear at high R ratios.

1.3 *Crack Arrest*

1.3.1 Crack Arrest of Microstructurally Short Cracks

The authors have already pointed to the importance of crack arrest for fatigue life and strength considerations. It is immediately evident that only cracks which do not arrest, negatively affect structural integrity. Crack arrest can happen at any of the crack propagation stages.

As mentioned above, *microstructurally short* cracks arrest at microstructural barriers, usually grain boundaries. It is important to understand that the fatigue limit is not a critical stress for crack initiation but a critical stress for avoiding or overcoming crack arrest (Miller 1993; Murakami 2002). With respect to the material fatigue limit (determined with smooth, unnotched specimens) it is that stress below which none of the numerous microstructural *short* cracks will grow permanently. Note, however, that in real (usually notched) structures other arrest mechanisms will dominate as will be discussed below. Note further that the material fatigue limit will disappear when mechanisms such as corrosion as a time dependent phenomenon are active which will overcome the crack arrest barriers.

1.3.2 Crack Arrest of Physically Short Cracks

Crack arrest of physically short cracks occurs due to the gradual build-up of the crack closure effects when the increase of ΔK with crack depth is smaller than the decrease of ΔK_{eff} due to an increasing K_{op}.

1.3.3 Crack Arrest of Long Cracks

Long cracks arrest when the applied ΔK is below the fatigue crack propagation threshold ΔK_{th}.

1.3.4 Crack Arrest of Short Cracks at Notches

This is probably the most common crack arrest scenario in components (Murakami 2002, 2003; Tanaka 2003). Due to the stress gradient away from the notch the rate of increase in ΔK becomes smaller, an effect which is additionally superimposed by the gradual build-up of the crack closure phenomenon. When the resultant ΔK_{eff} falls below the fatigue crack propagation threshold ΔK_{th}, crack arrest will occur.

1.3.5 Further Crack Arrest Scenarios

Further crack arrest cases are possible. Three options are: (i) Crack arrest can occur when the crack grows into a stress shielded area, e.g., near a geometrical stiffener as it is typical for thin-wall structures. (ii) Arrest also occurs when the crack grows towards an area of compressive residual stresses (e.g., in deep-rolled crankshafts) (iii) It also can happen when the crack propagates into a material with a higher threshold ΔK_{th} than that from where it started. An example is crack transition from a fine to a coarse grain zone, e.g. a heat affected zone in a weld. Note that the coarser grain is usually associated with a higher threshold (Plekhov et al. 2011). A plausible explanation for this is a stronger roughness-induced crack closure effect when the larger grain is associated with rougher crack faces, see Sect. 1.2.3.

Whilst, as mentioned above, the crack arrest mechanism of the microstructurally *short* cracks (Sect. 1.3.1) determines the material's fatigue limit, the component fatigue limit is usually controlled by the other mechanisms.

1.4 Describing Fatigue Crack Propagation and Arrest

1.4.1 Long Fatigue Crack Propagation (da/dN–ΔK Curve)

Long fatigue crack propagation is described by the da/dN–ΔK curve which is a sigmoidal curve in a log-log plot, Fig. 5.

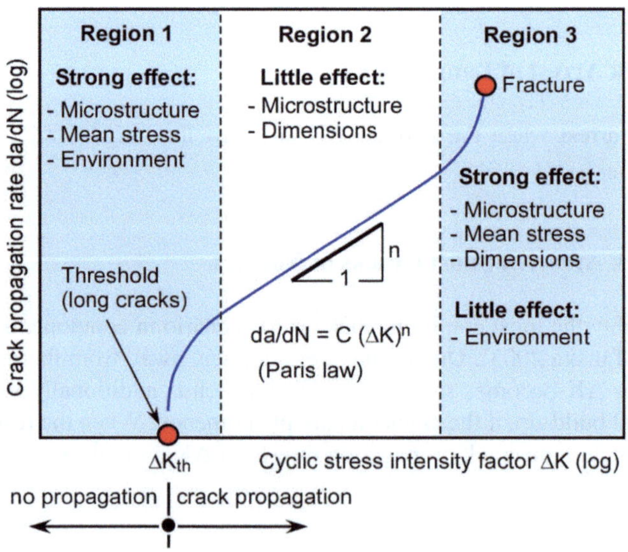

Fig. 5 Crack growth diagram da/dN–ΔK for long fatigue crack propagation; schematic view

The diagram can be subdivided into three regions:

(a) **Region 2 of the da/dN–ΔK Curve (Paris Regime)**

The central part of the double-logarithmic da/dN–ΔK curve (*Region* 2) is represented by a straight line, the so-called Paris law

$$da/dN = C(R) \cdot \Delta K^n = C(R) \cdot (K_{max} - K_{min})^n \tag{2}$$

or, when crack closure-corrected,

$$da/dN = C \cdot \Delta K_{eff}^n = C \cdot \left(K_{max} - K_{op}\right)^n. \tag{3}$$

The parameter C is a constant which besides the material depends on the loading conditions, particularly the mean load $\bar{K} = 0.5\ (K_{max} + K_{min})$ or the load ratio R ($= K_{min}/K_{max}$), but also on the environment. The exponent n is the slope of the curve in double-logarithmic scaling, and it is usually in a range from n = 2 to 4 for metallic materials (Ritchie 1999).

(b) **Region 1 of the da/dN–ΔK curve (Threshold Regime)**

In *Region* 1, the crack growth diminishes rapidly with decreasing ΔK until it reaches a threshold value ΔK_{th} below which crack arrest occurs. Whilst the treatment of the crack propagation in the Paris regime is state-of-the-art today (a statement that, however, has to be put in perspective for variable amplitude loading), the situation is much less clear with respect to the threshold regime (keyword: similitude, see Tanaka 2003; Schijve 2004; also Zerbst et al. 2016b).

(c) **Region 3 of the da/dN–ΔK Curve (Transition to Fracture Regime)**

Crack acceleration is stated in Region 3 due to interaction events between cyclic and monotonic crack growth mechanisms. Finally, crack propagation is terminated by monotonic fracture when the maximum stress intensity in the loading cycle, K_{max}, approaches the monotonic fracture resistance. Note that *Region* 3 plays a minor role with respect to fatigue lifetime. Because of its high crack growth rate it refers to only a small portion of this. No special attention is given to it in IBESS for that reason.

With respect to the curve fit of the da/dN–ΔK curve, IBESS follows the NASGRO approach (NASGRO 2000) based on Forman and Mettu (1992):

$$\frac{da}{dN} = C \cdot [U \cdot \Delta K]^n \cdot \left[1 - \frac{\Delta K_{th}}{\Delta K}\right]^p \Big/ \left[1 - \frac{K_{max}}{K_{mat}}\right]^q, \tag{4}$$

however, with a number of modifications which will be discussed in detail in Sect. 2.2.4.2. In Eq. (4) the term U is a crack closure parameter, ΔK_{th} the fatigue crack propagation threshold the determination of which will be discussed in Sects. 2.2.4.1 and 2.2.4.4 and K_{mat} stands for the monotonic fracture resistance. The term $[1 - \Delta K_{th}/\Delta K]^p$—p is a fit parameter—describes the threshold range (Region 1) of the da/dN–ΔK curve in that da/dN is zero for $\Delta K_{th} \geq \Delta K_p$. The *Region* 3 term $\left(1 - K_{max} / K_{mat}\right)^q$ is skipped in IBESS for the reason mentioned above.

1.4.2 Long Fatigue Crack Propagation Threshold $\Delta K_{th,LC}$

The fatigue crack propagation threshold consists of two components, an intrinsic one, $\Delta K_{th,eff}$, which is a material parameter, and an extrinsic one, $\Delta K_{th,op}$, related to the crack closure phenomenon, see Sect. 1.2.

$$\Delta K_{th} = \Delta K_{th,eff} + \Delta K_{th,op} \tag{5}$$

The intrinsic component $\Delta K_{th,eff}$ is identical to the closure-free ΔK_{th}, e.g. determined at high R-ratios, typically at $R \geq 0.8$; therefore, it is designated with the index "eff". It is correlated with the modulus of elasticity, E, which in turn is a function of the crystal lattice of the material, and by the Burger's vector, $\|b\|$, as a geometrical measure of dislocations, see, Sect. 2.3.2.6. Typical values of $\Delta K_{th,eff}$ are in the order of 2.4–2.6 MPa $m^{1/2}$ for steels, 0.9–1.9 MPa $m^{1/2}$ for aluminium alloys and 1.7–2.5 MPa $m^{1/2}$ for titanium alloys (Hadrboletz et al. 1994, also for a more extended overview). Whilst $\Delta K_{th,eff}$ is independent of material parameters related to the microstructure as well as of the R-ratio and the crack length, the crack closure component $\Delta K_{th,op}$ is affected by all these parameters, see the discussion by Zerbst et al. (2016b).

Some remarks are due with respect to the experimental determination of ΔK_{th}, for a more detailed discussion see again Zerbst et al. (2016b). The common method according to test standards such as ASTM E 647 (2011) or ISO 12108 (2012) is stepwise load decrease (Fig. 6) until the crack propagation rate falls below a certain value.

Note that this limit value is different for the two standards. ASTM E 647 sets it to $(da/dN)_{th} = 10^{-7}$ mm/cycle whereas ISO 12108 specifies it at $(da/dN)_{th} = 10^{-8}$ mm/cycle. That this difference might be a problem in practical application is illustrated in Fig. 7 where the da/dN–ΔK curve changes its slope below da/dN$=10^{-7}$ mm/cycle. Note further that crack propagation rates below this order are faced with the fundamental problem that there must be a discrete minimum growth step per cycle associated with the lattice arrangement of the atoms in crystals and the minimum displacement provided by the Burgers vector, which, for steels, is in the order of 10^{-7} mm.

A second problem is associated with the oxide debris-induced crack closure effect in materials susceptible to corrosion. Figure 8 illustrates the effect of the gradient of load reduction on the threshold. Reading the curve from right to left, ΔK_{th} decreases down to a minimum value. This minimum defines the gradient in the test standards not to be exceeded when premature (plasticity-induced) crack closure shall be avoided. However, as discussed in Sect. 1.2, the plasticity-induced one is not the only crack closure effect. The oxide debris-induced effect is promoted by a small load gradient which corresponds to a larger time span available for corrosion which is generally known to be a time-dependent phenomenon. As the consequence, the

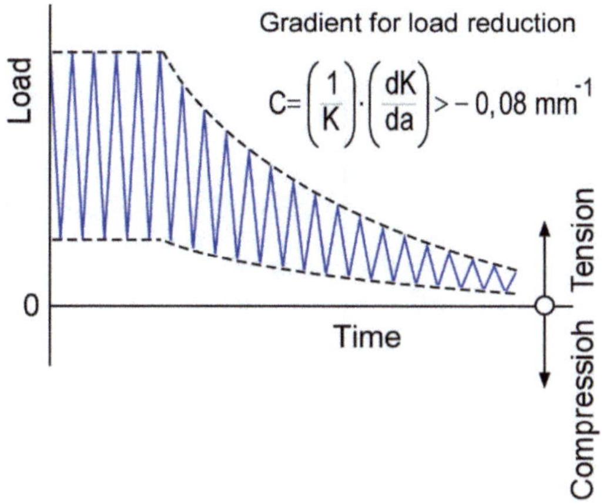

Fig. 6 Determination of the fatigue crack propagation threshold: load reduction technique according to ASTM E 647 (2011)

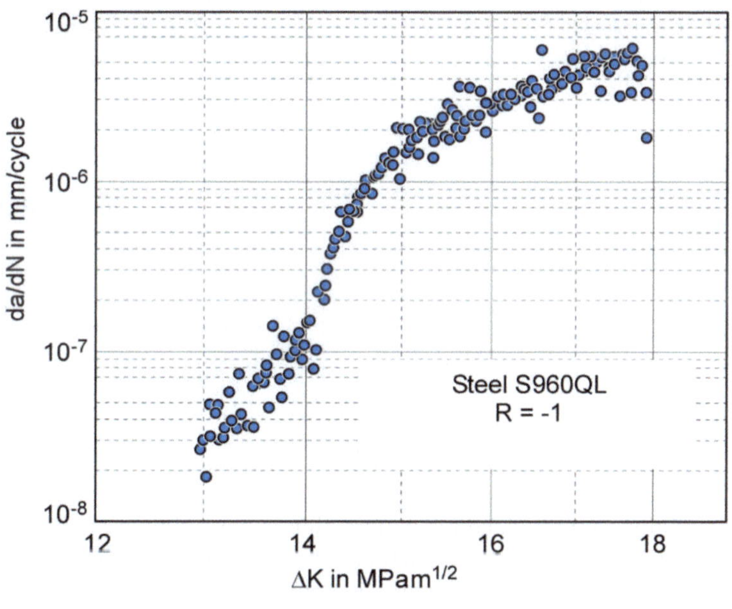

Fig. 7 Crack propagation curves for higher strength steel S960QL obtained in ambient air; according to Madia and Zerbst (see Zerbst et al. 2016b)

threshold will increase again in corrosion-prone materials when the loading gradient becomes smaller. The result can be an experimental threshold larger than it would be under inert conditions or under conditions which did not allow for the crack closure phenomenon to develop.

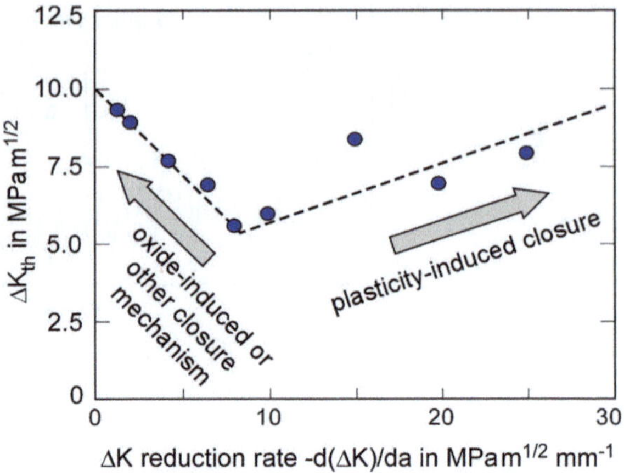

Fig. 8 Effect of the load reduction rate on the fatigue crack propagation threshold ΔK_{th}; according to Brook (1983); arrows and interpretation are added by the authors

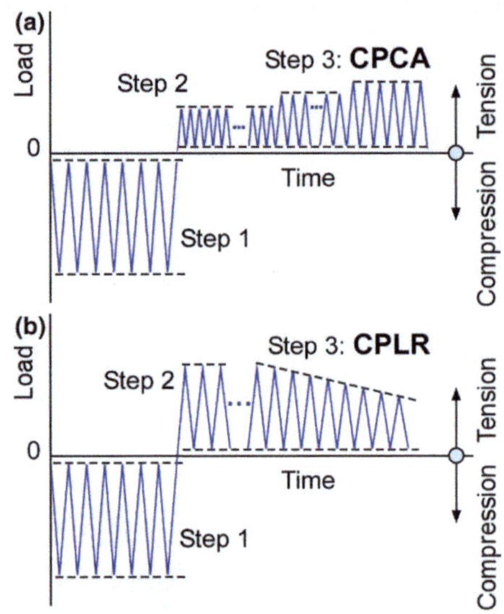

Fig. 9 Determination of the fatigue crack propagation threshold: **a** compression pre-cracking constant amplitude (CPCA) method; **b** compression pre-cracking load reduction (CPLR) method

Such conditions are provided by compression pre-cracking tests. Figure 9 illustrates two possible schemes of this method when applied to *long* crack thresholds. The acronym CPCA stands for "compression pre-cracking constant amplitude", the acronym CPLR for "compression pre-cracking load reduction". Guidance on the methods is, e.g., provided by Newman et al. (2005, 2010). That compression pre-cracking may yield lower fatigue crack propagation thresholds than the load reduction method of the common test standards is illustrated in Fig. 10.

Note that the intrinsic threshold $\Delta K_{th,eff}$ is not affected by the testing procedure. It can be experimentally determined by compression pre-cracking, too, but—easier—by a K_{max}-method (where the R ratio increases during the test) such as illustrated in Fig. 11 or by load reduction at high R ratio.

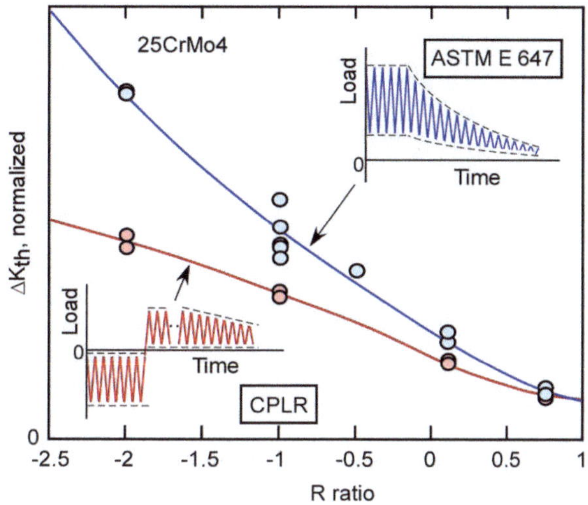

Fig. 10 R ratio dependency of fatigue crack propagation threshold values obtained by different experimental methods, quenched and tempered steel; according to Carboni and Regazzi (2011)

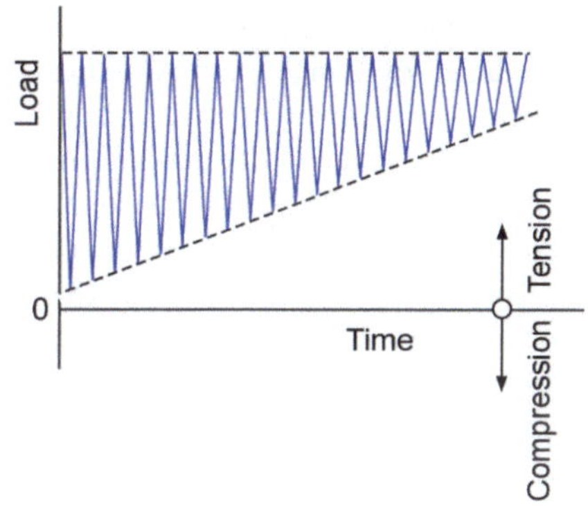

Fig. 11 Determination of the fatigue crack propagation threshold: Constant K_{max} method

1.4.3 Physically Short Crack Propagation and Propagation Threshold ΔK_{th} (a) (Cyclic R-curve)

Whilst the intrinsic threshold $\Delta K_{th,eff}$ is a material parameter, the crack closure or extrinsic component of ΔK_{th}, $\Delta K_{th,op}$, depends on parameters such as the load ratio, the roughness of the crack faces in the crack wake, the cyclic deformation behaviour and the crack length. At the *physically short* crack stage $\Delta K_{th,op}$ is zero at the beginning but it gradually increases during crack propagation, see Sect. 1.1.3. This scenario is schematically illustrated in Fig. 12a. At zero crack growth, $\Delta a = 0$, the threshold $\Delta K_{th} = \Delta K_{th,eff}$. Then it increases due to an enlarging $\Delta K_{th,op}$ component until the *long* crack stage is reached and $\Delta K_{th,op}$ and ΔK_{th} become crack size independent. Note that the referring crack size as well as the overall threshold ΔK_{th} are influenced by all crack closure mechanisms (Fig. 12b) because of which quantitatively different pictures can emerge, e.g., under different environmental conditions. The crack

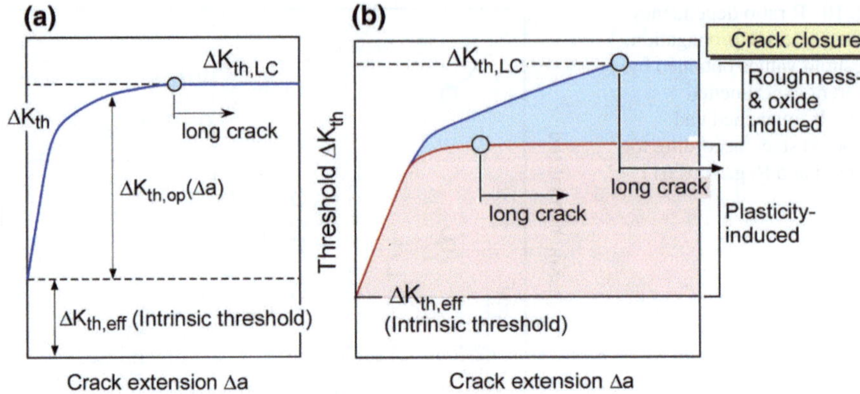

Fig. 12 Cyclic R-curve. **a** Principle; **b** schematic build-up of the different crack closure mechanisms; according to Maierhofer et al. (2018)

size dependency of ΔK_{th} is designated as "cyclic R-curve", a term which was first introduced by Tanaka and Akinawa (1988) in the context of crack propagation at notches.

The experimental determination of the cyclic R-curve, as it was also performed in the frame of IBESS, follows a methodology developed by the group of Pippan at Leoben (e.g., Tabernig and Pippan 2002), for a detailed discussion see Zerbst et al. (2016b) and Maierhofer et al. (2018). The method is illustrated in Fig. 13. Subsequent to compression pre-cracking, ΔK is stepwise increased keeping constant the load ratio R of interest. As long as $\Delta K \leq \Delta K_{th,eff}$, nothing happens. In the range $\Delta K_{th,eff} < \Delta K \leq \Delta K_{th,LC}$ the crack starts to propagate but after some extension it arrests due to the gradual build-up of the crack closure effects which reduce ΔK_{eff}. When, finally, the *long* crack threshold $\Delta K_{th,LC}$ is exceeded, no arrest occurs any longer. The cyclic R-curve is obtained by connecting the arrest points (Δa, ΔK).

The experiments for cyclic R-curve determination are usually performed by a single-specimen technique with the crack depth being monitored during the test by the potential drop method. Note that, due to the very slow crack propagation (a test may last for days or even weeks) temperature compensation is necessary. In order to be certain to exclude time-dependent corrosion effects, Zerbst and Madia (2014b), in addition, carried out multiple specimen tests which realized just one load step per specimen. Thus, a number of specimens was needed for generating one cyclic R-curve. The crack extension was measured post mortem at the fracture surfaces. A comparison between both techniques is shown for one of the steels investigated within IBESS, S355NL base metal, which did not show susceptibility to atmospheric corrosion. As expected, both curves coincided in this case (Fig. 14).

An alternative method for determining the cyclic R-curve is provided by a modified El Haddad fit of the Kitagawa-Takahashi diagram (Zerbst and Madia 2014b). With the fit parameter a_0, determined for a semi-circular crack in a tension loaded plate, i.e., the geometry function $Y = 0.728$ (Tanaka and Akinawa 2003),

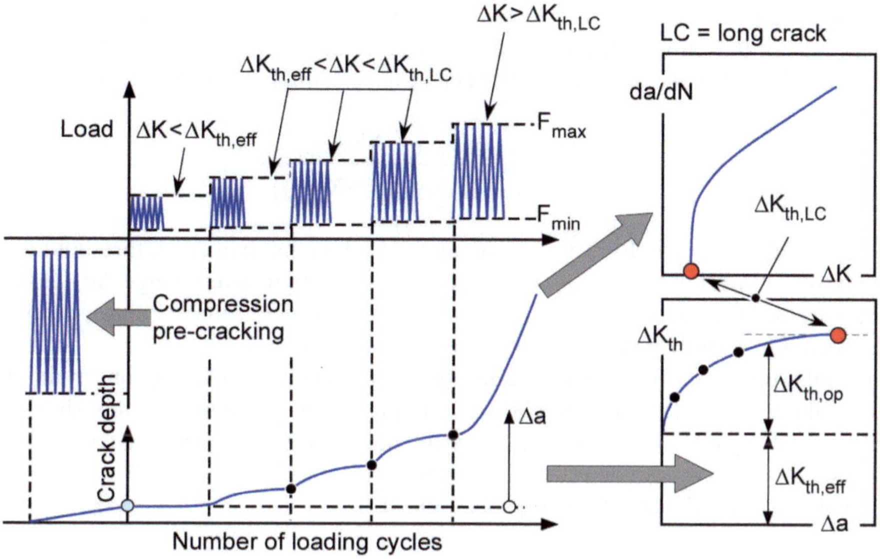

Fig. 13 Loading scheme of cyclic R-curve determination according to Tabernig and Pippan (2002)

Fig. 14 Comparison of cyclic R-curves obtained by single and multiple specimen methods for S355NL steel; according to Zerbst and Madia (2014b)

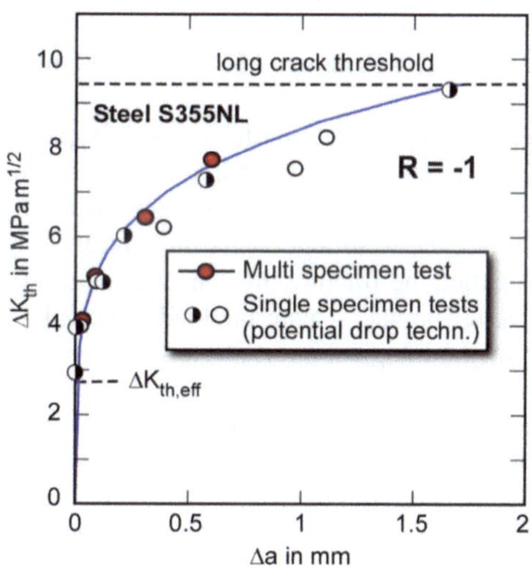

$$a_0 = \frac{1}{\pi}\left(\frac{\Delta K_{th,LC}}{Y \cdot \Delta\sigma_e}\right)^2 \tag{6}$$

the cyclic R-curve is obtained by

$$\Delta K_{th}(\Delta a) = \Delta K_{th,LC} \cdot \sqrt{\frac{\Delta a + a^*}{\Delta a + a^* + a_0}}. \tag{7}$$

The additional term a* is necessary to fulfil the condition $\Delta K_{th}(\Delta a = 0) = \Delta K_{th,eff}$, i.e. the minimum threshold is not zero but the intrinsic one (Vormwald et al. 1992; Zerbst and Madia 2014b). The term can simply be determined by

$$a* = a_0 \cdot \frac{\left(\Delta K_{th,eff} / \Delta K_{th,LC}\right)^2}{1 - \left(\Delta K_{th,eff} / \Delta K_{th,LC}\right)^2} \tag{8}$$

This way, the cyclic R-curve is estimated based on the *long* crack threshold and the intrinsic threshold, $\Delta K_{th,LC}$ and $\Delta K_{th,eff}$, and the fatigue stress range $\Delta\sigma_e$. Some caution is required with respect to literature data for a_0 which, without being explicitly stated, are frequently given for $Y = 1$, see Sect. 2.3.2.6.

Whilst this is a simple task, a much bigger problem is caused by uncertainties in the input parameters $\Delta\sigma e$ and $\Delta K_{th,LC}$. In particular, the latter not only shows a large scatter (Sect. 2.3.2.5.2) but also a systematic effect of the experimental method applied to its determination (cf. Fig. 10 in Sect. 1.4.2).

Note that a_0 is dependent on the load ratio R.

1.5 Fatigue Life and Fracture Mechanics Triangle

1.5.1 The Fracture Mechanics Triangle

Figure 15 provides an illustration of typical weld defects as they occur in manual arc (stick) welding of mild steels. This overview is provided from the viewpoint of failure analysis. Note that some authors even count residual stresses as well as misalignment among the welding defects. In contrast, a fracture mechanics analysis follows what is called the fracture mechanics triangle. In general terms this consists of the three "vertex points"—material, load and crack—which have to be separately treated from a methodological point of view. This is, however, difficult in the context of fracture mechanics application to fatigue life. The corresponding fracture mechanics triangle with special emphasis on weldments is schematically outlined in Fig. 16. Note that the elements (vertices), when applied to the early crack growth stages, are not strictly independent but interplay, e.g., with respect to defects which can be interpreted as constituent elements of the material or as crack-like defects.

1.5.2 Material

The material "vertex point" includes basic material characteristics such as the microstructure but also flaws such as inclusions or even slag when embedded in the material. However, in a fracture mechanics analysis these are not explicitly considered but implicitly in that they influence the phenomenological properties da/dN–ΔK curve, fatigue crack propagation threshold ΔK_{th}, monotonic fracture resistance, etc.

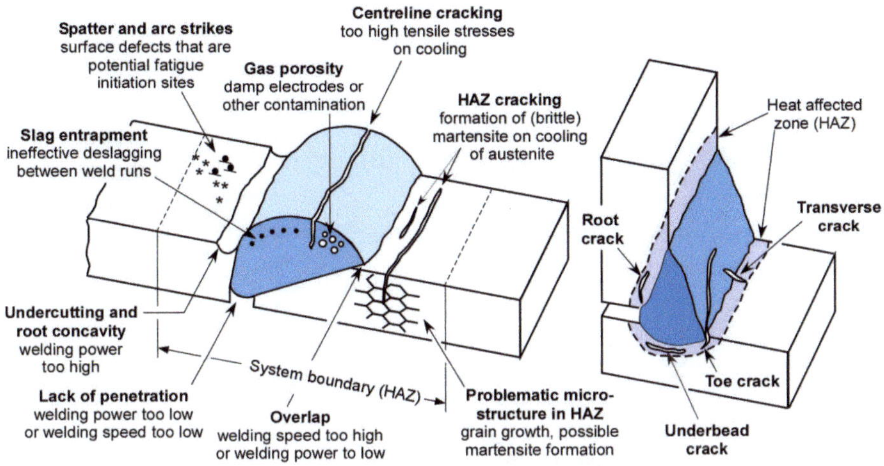

Fig. 15 Potential welding defects of manual arc (stick) welding of typical mild steels from the point of view of a failure analyst; according to Gagg (2005)

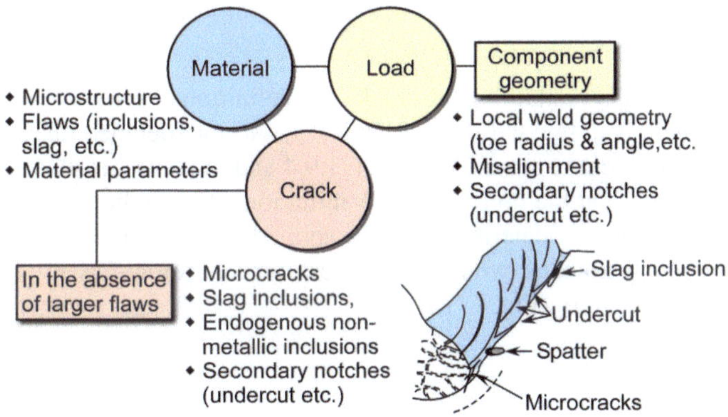

Fig. 16 Fracture mechanics triangle in the context of fracture mechanics-based fatigue strength determination. As can be seen, some of the features could be assigned to different "vertex points"

1.5.3 Load

With respect to loading, an (analytical) fracture mechanics analysis distinguishes between *primary* and *secondary* stresses, σ^p and σ^s. *Primary* stresses are caused by applied mechanical loads, *secondary* stresses by internal strain mis-match. Note that typical *secondary* stresses are welding residual stresses but, vice versa, welding residual stresses can be both, *primary* or *secondary*. The decisive characteristics of a *secondary* stress are that the forces and moments behind them self-equilibrate across

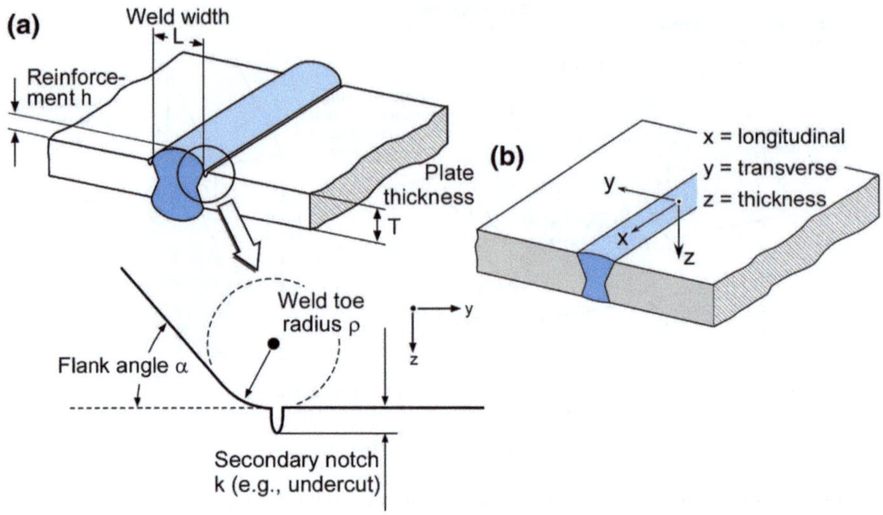

Fig. 17 **a** Definition of the parameters of the weld toe geometry; **b** coordinate system applied to the weldment

the section potentially containing the crack. The definition of *primary* and *secondary* stresses used here is in line with BS 7910 (2013). A more detailed discussion of the residual stress problem is provided in Sect. 1.9.4 and in Hensel et al. (2018).

The local stresses at the potential crack initiation sites (in the present case along the weld toe) are not only determined by the applied net section stress but also by the weld geometry, e.g. notches, which cause stress concentration. Essential parameters of the weld toe geometry (Fig. 17) are

- the weld toe radius ρ,
- the flank angle α,
- the weld reinforcement h,
- the weld width L, and
- secondary notches of depth k.

Secondary notches can be undercuts, roughness grooves of the base plate adjacent to the weld toe or edge points of welding ripples.

1.5.4 Crack

In the absence of large defects such as lack of penetration, flaws can be microcracks, e.g., next to spatter points, where they are caused by thermal stresses during spatter cooling, but also inclusion clusters or secondary notches such as undercuts. The two latter items again illustrate what the authors had in mind when they noted that the separate treatment of the triangle "vertex points" can be a problem in the present context. Note that inclusion clusters may also affect the material properties and secondary notches can act as stress concentrators at the "load vertex" but also be part of an initial crack when their notch tip radius is smaller than the depth of the crack emanating from them, for a more detailed discussion see Sect. 2.3.1.5.

What further complicates the matter is that only such cracks which are not arrested are initial cracks in a fatigue life analysis, see Sect. 1.3. This brings up the next point: When is an initial flaw an initial crack in a fracture mechanics context?

1.6 Initial Crack Size for Fracture Mechanics Analysis

A problem on its own is the definition of the initial crack dimensions for fracture mechanics analyses, for a detailed discussion of this topic see Zerbst et al. (2018c). In the context of damage tolerant design the initial crack size is usually provided by the detection limit of non-destructive testing. This is, however, not possible with respect to the total lifetime of an S-N curve analysis where the initial crack size is usually far below this detection limit.

At this place, a remark regarding the current practice is due. A widely applied concept for specifying an initial crack size in the context of fatigue life and strength is the "equivalent initial flaw size" (EIFS) approach, for a brief review see Johnson (2010). The initial crack size is recalculated from or adapted to S-N data. The calculations are usually based on the linear elastic ΔK concept, i.e., the *long* crack da/dN–ΔK curve and the propagation of just one crack is assumed. Crack closure and residual stress effects, if at all, are considered in a very simplified manner. In documents such as the IIW guidelines (Hobbacher 2016) the results of such analyses are generalized to a fixed initial crack depth and geometry, e.g. a crack depth of 100 μm or more is suggested to be used for linear elastic fatigue crack propagation analyses. For more detailed overviews of such fixed crack depths and aspect ratios a/c see Radaj et al. (2006) and Schork et al. (2018). It is clear, that this initial crack size is a model parameter rather than a physically meaningful quantity. It reaches its limits when transferability problems (e.g., between different component geometries) come into play.

To avoid these shortcomings, the IBESS approach follows a different philosophy which will be explained in detail in Sect. 2.4.5. Here it shall only be mentioned that a lower bound initial crack size is determined by a crack arrest analysis in conjunction with the fatigue limit of the material. If material (or welding) defects larger than this minimum value exist, these take over the role of the initial crack size, i.e., IBESS uses a two-parameter concept for defining the parameter.

1.7 Multiple Crack Propagation

When the fatigue limit was defined above as that stress above which just one crack is capable to grow, it must be added now, that there are many applications (including welds) where multiple crack propagation occurs at load levels higher than this. As a rule, the number of propagating cracks increases with the stress level (Lecsec et al. 1995; Schork et al. 2018). An example which was obtained within the IBESS project is shown in Fig. 18 where cracks along the weld toe are visualized by heat tinting after 1/4–1/3 of the overall lifetime known from S-N testing.

The multiple crack propagation phenomenon (Fig. 19a) is influenced by the geometrical irregularity of the weld toe and it has an effect on the crack aspect ratio a/c during fatigue (Fig. 19c). Note that the latter abruptly changes due to successive coalescence events of individual cracks. There might, however, be exceptions from that rule in the presence of very large initial defects which not only dominate crack initiation but also subsequent propagation.

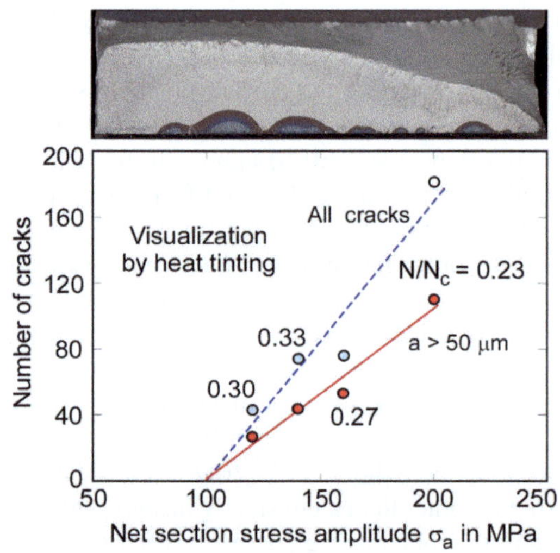

Fig. 18 Experimentally determined number of cracks along the 50 mm weld toe of butt welds, obtained at 1/4–1/3 of the overall lifetime; according to Schork et al. (2018), see also Zerbst et al. (2016a)

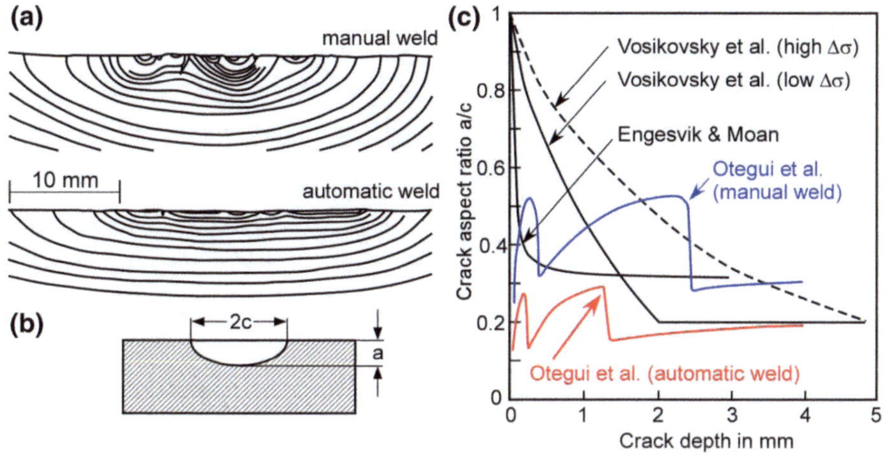

Fig. 19 Development of the crack shape during multiple crack propagation. **a** Different patterns of a manual and an automatic weld; **b** definition of crack depth and length at surface; **c** comparison of these results with frequently used literature equations for a/c development which are usually based on the assumption of one instead of multiple cracks; according to Otegui et al. (1991), for more detailed discussion see also Schork et al. (2018)

1.8 The Need of Probabilistic Assessment

As mentioned above, multiple crack propagation of cracks originating along a weld toe is a consequence of the non-uniform geometry of the latter along its length dimension. An example from the IBESS project is provided in Fig. 20. It is immediately evident that any multiple crack propagation analysis must consider the statistical variability of the input parameters. The principle, how this is realized within the IBESS approach, is described in detail in Sect. 2.4.2.

1.9 Factors Affecting Monotonic Fracture and Fatigue Crack Propagation of Welds

1.9.1 Brief Overview

Besides geometrical aspects (the exact local weld geometry is usually unknown), weldments are characterized by at least three specific features:

(a) Pronounced inhomogeneity of the microstructure across the weld seam and the adjacent base metal,
(b) Strength mis-match between the different zones of the weld (not always) as one consequence of (a), and
(c) Welding residual stresses.

Fig. 20 Variation of the weld toe radius and the flank angle along the toe of a butt weld of steel S355NL; according to Schork et al. (2018)

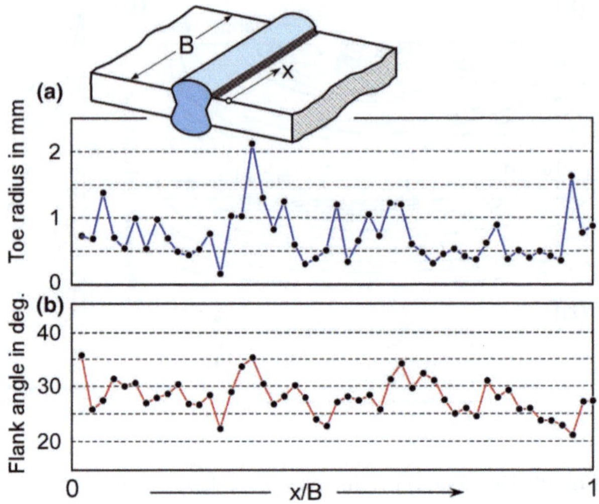

A detailed discussion of the consequences of these issues on fracture mechanics material testing and component assessment (predominantly for monotonic loading) is provided by Zerbst et al. (2014a).

1.9.2 Inhomogeneous Microstructure and Material Properties

1.9.2.1 Inhomogeneous Microstructure

The inhomogeneous microstructure is the consequence of different cooling rates at different positions in and adjacent to the weld seam such that the various positions experience different paths in the temperature-time-transition (TTT) diagram. In multi-pass welds, local re-heating occurs which additionally causes temper effects. Particularly the heat affected zone (HAZ) adjacent to the molten area shows a wide variety of quite different microstructures which correspond to different strength, ductility and toughness properties. An example is provided in Fig. 21. where the author identified as many as six different microstructures exclusively in the HAZ of a multi-pass weld of low alloy carbon-manganese steel.

Two examples taken from IBESS are provided in Figs. 22 and 23, where the microstructures of base metal and HAZ and the hardness distributions across the welds of two steels of significantly different strengths are shown. Whilst the base metals are characterized by ferritic-pearlitic (S355NL steel) and martensitic (S960QL steel) microstructures, the HAZ microstructure of both materials was martensitic-bainitic. In addition, there was a difference between the two cap passes which experienced different welding speeds (30 and 35 cm/min) in combination with different energy inputs (1200 and 1000 J/mm).

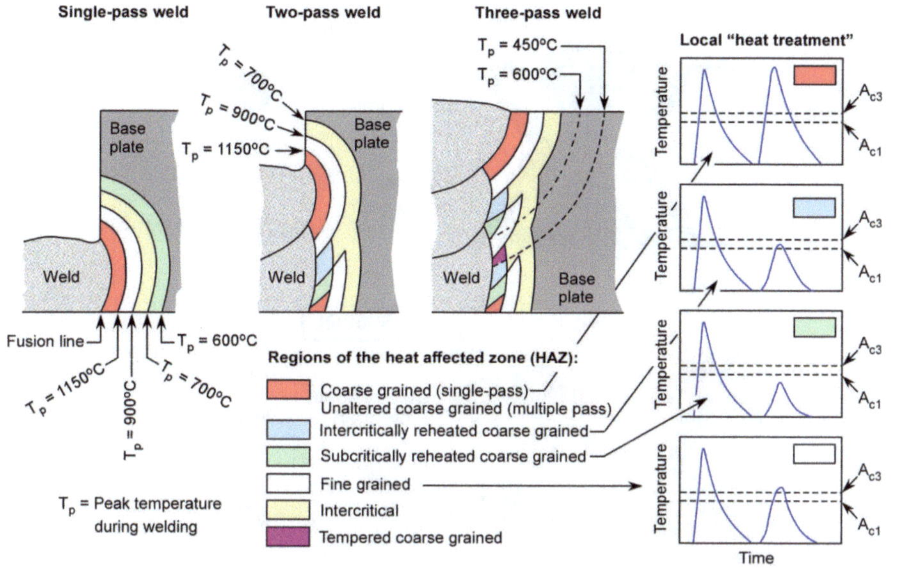

Fig. 21 Schematic representation of the heat affected zone (HAZ) microstructure of a carbon-manganese steel weld; from left to right: single-pass, two-pass and three-pass weld; according to Toyoda (1989)

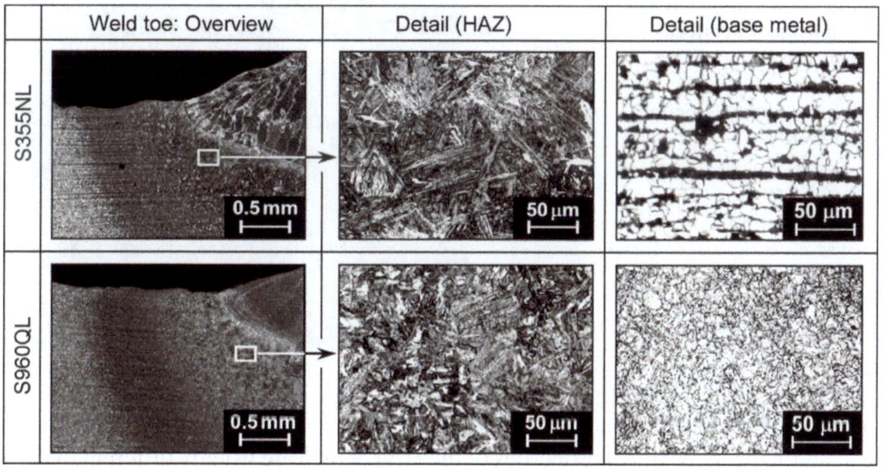

Fig. 22 Metallographic views of the microstructures near the weld toes for welded joints made of steels S355NL and S960QL which have been investigated in IBESS; according to Kucharczyk et al. (2018)

It is important for fatigue considerations that the crack usually initiates at the weld toe and then propagates into a different material zone where it spends most of the total lifetime of the weld. In Fig. 24, also taken from IBESS, this material zone is

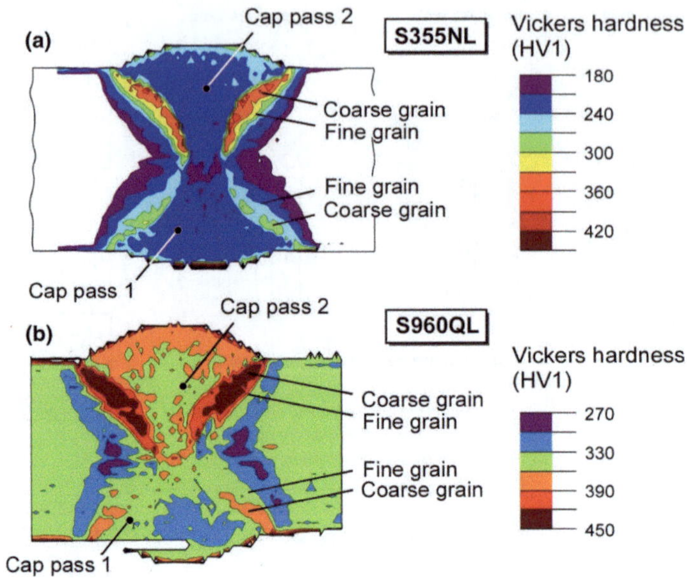

Fig. 23 Hardness distributions across the butt welds made of 10 mm thick plates of steel S355NL and S960QL which have been investigated in IBESS; according to Kucharczyk et al. (2018)

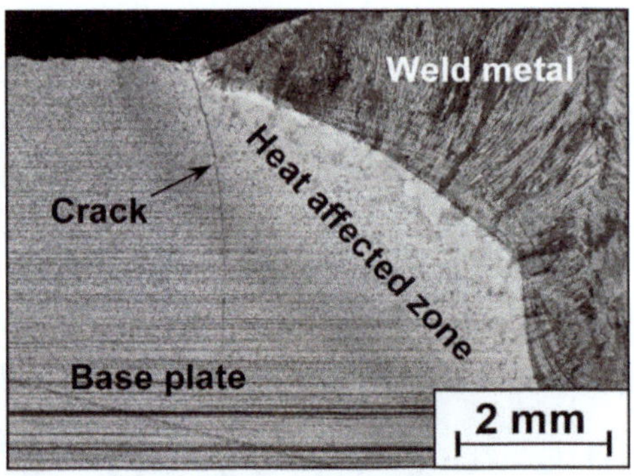

Fig. 24 Crack initiation and growth in the heat affected zone in a butt weld, Steel S355NL investigated in IBESS; according to Schork et al. (2018)

the HAZ. Therefore, it is most essential to base the (*short*) fatigue crack propagation analysis on the mechanical properties of the HAZ which are usually different to those of the base metal. Note that, within the IBESS project, all HAZ material data have been obtained on test specimens consisting of thermally simulated microstructure, see also Sect. 2.3.2. The technique performed for this will not be discussed here in detail, however, some guidance is provided by Kucharczyk et al. (2018).

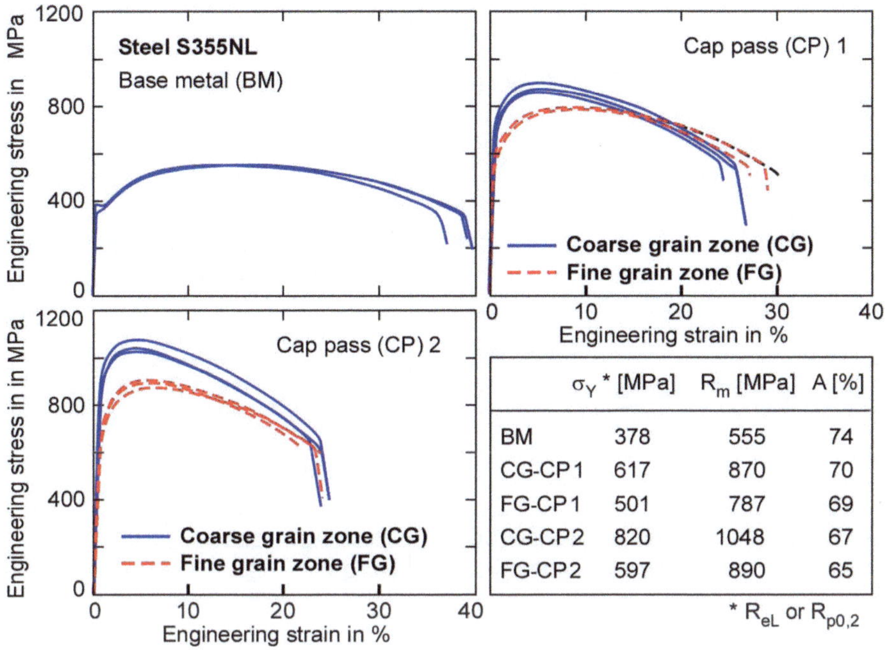

	σ_Y * [MPa]	R_m [MPa]	A [%]
BM	378	555	74
CG-CP1	617	870	70
FG-CP1	501	787	69
CG-CP2	820	1048	67
FG-CP2	597	890	65

* R_{eL} or $R_{p0,2}$

Fig. 25 Monotonic stress-strain curves of different microstructures and cap passes of steel S355NL, cf. Fig. 23; according to Kucharczyk et al. (2018)

1.9.2.2 Monotonic and Cyclic Stress-Strain Data

The different microstructures are correlated with different strength and deformation properties. Figures 25, 26 and 27 show the corresponding monotonic and—in the present context more important—(stabilized) cyclic stress-strain curves of the two steels under consideration. Besides the base metal, the properties of different zones across the HAZ have been characterized. For more detailed information see Kucharczyk et al. (2018).

1.9.2.3 Monotonic Fracture Toughness

An important effect of the different microstructures on the monotonic fracture resistance was that the base metal specimens of the two steels failed on the upper shelf whereas the HAZ specimens failed in the ductile-to-brittle transition range of the fracture toughness-temperature curve (for the definition of these terms see Fig. 28). The results are shown in Figs. 29 and 30.

Two remarks are due with respect to Figs. 29b and 30b. The determination of the cumulative probability function $P_f(K_c^J)$ of the three-parameter Weibull distribu-

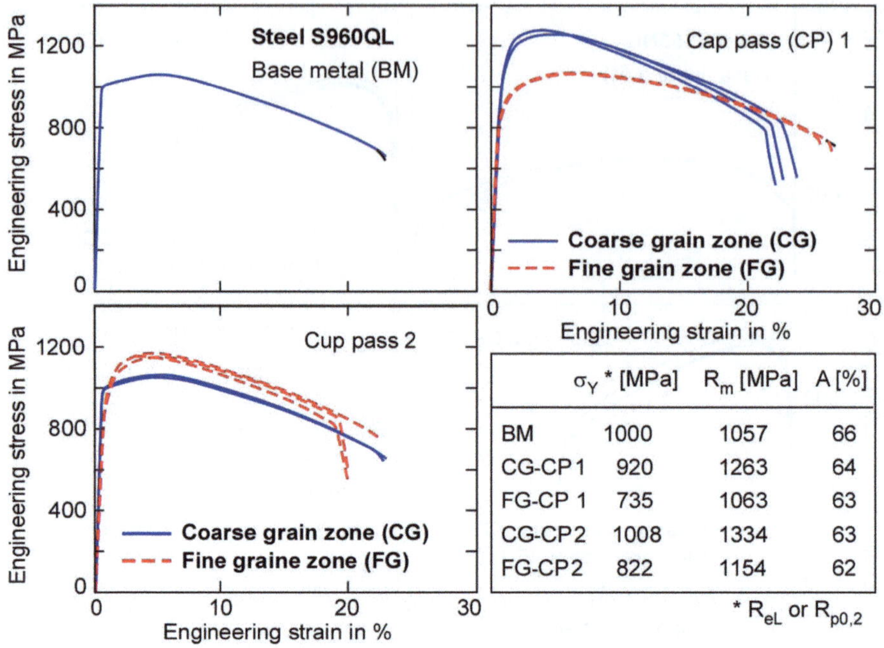

	σ_Y * [MPa]	R_m [MPa]	A [%]
BM	1000	1057	66
CG-CP1	920	1263	64
FG-CP 1	735	1063	63
CG-CP2	1008	1334	63
FG-CP2	822	1154	62

Fig. 26 Monotonic stress-strain curves of different microstructures and cap passes of steel S960QL, cf. Fig. 23; according to Kucharczyk et al. (2018)

tion was based on the ASTM E 1921 (2010) test standard although two different definitions of P were used such as shown in the legends. Note that the application of the guideline is restricted to ferritic steels with yield strengths ranging from 275 to 825 MPa and welds with strength mis-match smaller than 10% (for the definition of the latter, see Sect. 1.9.3). However, as mentioned above, the HAZ materials showed martensitic-bainitic microstructures and, largely, higher yield strengths (see Sect. 1.9.2.2). Therefore, the applicability had to be demonstrated on an empirical basis; in other words: it had to be checked whether the provided equation would reasonably fit the measuring points. This was very roughly the case in the present examples although the lower tail was underestimated. A more detailed discussion on fracture resistance determination of HAZ material is provided in Annex A.

1.9.3 Strength Mis-match

Strength mis-match is one of the consequences of the inhomogeneous microstructure across the weld. The term means that the different zones of the weld show different stress-strain curves such as illustrated in the example of Sect. 1.9.2. Note, however, that the phenomenon usually is defined in a simplified, less detailed way in that only

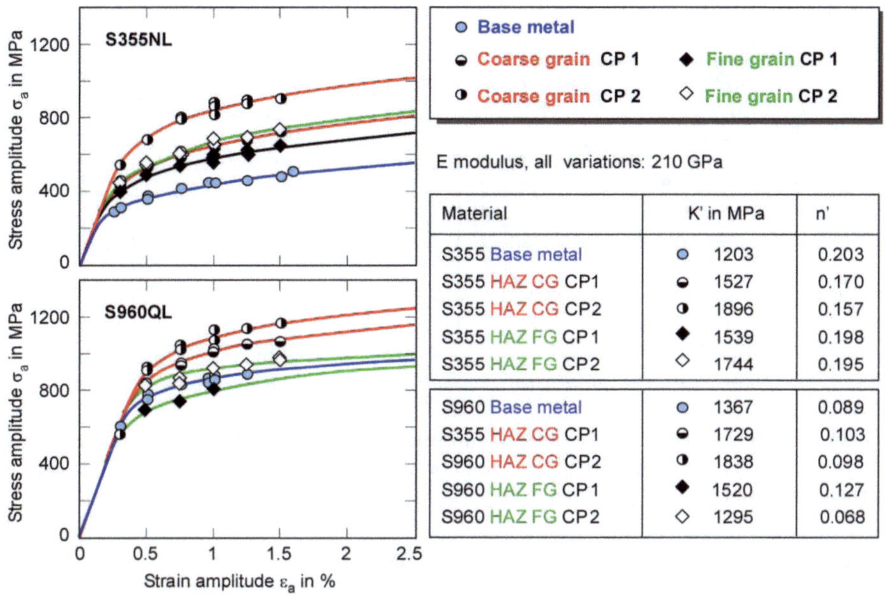

Material		K' in MPa	n'
S355 Base metal	○	1203	0.203
S355 HAZ CG CP1	◒	1527	0.170
S355 HAZ CG CP2	◑	1896	0.157
S355 HAZ FG CP1	◆	1539	0.198
S355 HAZ FG CP2	◇	1744	0.195
S960 Base metal	○	1367	0.089
S355 HAZ CG CP1	◒	1729	0.103
S960 HAZ CG CP2	◑	1838	0.098
S960 HAZ FG CP1	◆	1520	0.127
S960 HAZ FG CP2	◇	1295	0.068

Fig. 27 Stabilized cyclic stress-strain curves of different microstructures and cap passes of steels S355NL and S960QL, cf. Fig. 23; according to Kucharczyk et al. (2018)

Fig. 28 Dependency of the monotonic fracture toughness on the temperature of a bcc material; schematic view

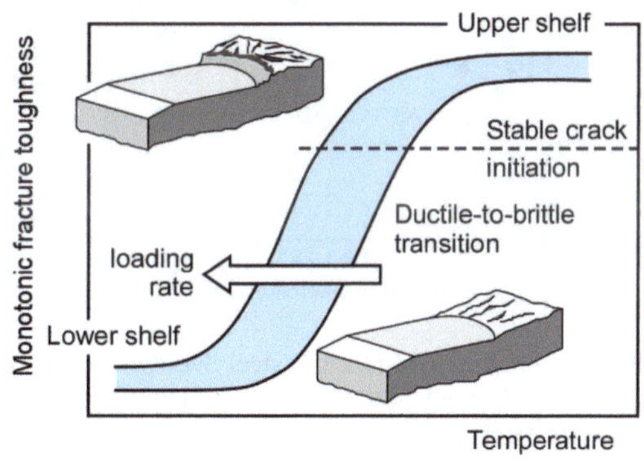

the yield strengths of the base and weld material, σ_{YB} and σ_{YW}, are considered by a strength mis-match factor M. Its common definition is given by

$$M = \sigma_{YW} / \sigma_{YB} \qquad (9)$$

with σ_{YW} and σ_{YB} being the yield strengths of the weld and base metal. The range $M > 1$ is designated as over-matching (OM), $M = 1$ as even-matching, and $M < 1$ as under-matching (UM). Usually, in order to shield the weld, moderate over-matching in the range between $M = 1.15–1.3$ is targeted but there are also cases of much

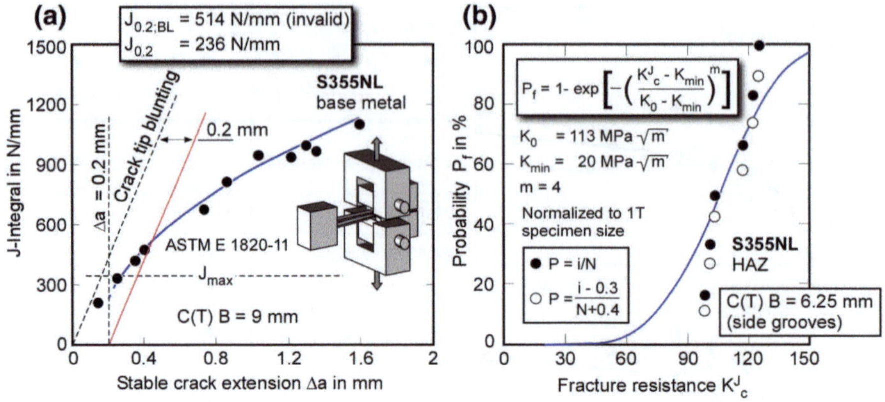

Fig. 29 Fracture resistance of steel S355NL investigated within IBESS; **a** base metal: resistance against stable crack initiation; **b** HAZ: ductile-to-brittle-transition fracture toughness; according to Kucharczyk et al. (2018)

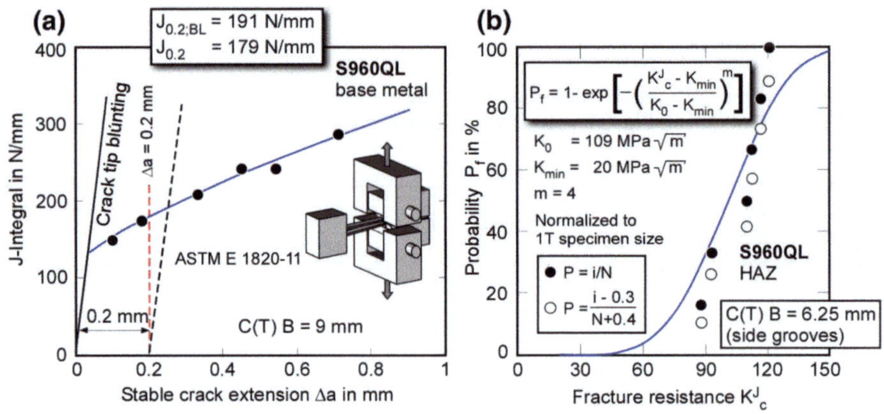

Fig. 30 Fracture resistance of steel S960QL investigated within IBESS; **a** base metal: resistance against stable crack initiation; **b** HAZ: ductile-to-brittle-transition fracture toughness; according to Kucharczyk et al. (2018)

higher values and also those of undermatching. Rather large and small M values, i.e. pronounced OM and UM, are typical for electron and laser beam welds. Two examples are shown in Fig. 31 where the hardness distributions, which correlate with the strength, are reproduced. Note that undermatching occurs in highly metastable materials such as aluminium alloys or high-strength steels with a high yield-to-tensile ratio.

What is the nature of the effect of strength mis-match on the mechanical behavior of a component? Imagine a plate consisting of two materials with different strengths. When tensile loaded, the less strong material will yield first and it will experience

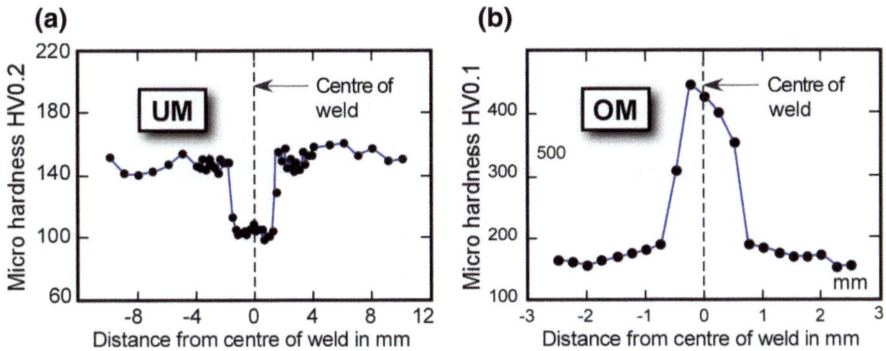

Fig. 31 Hardness profiles of (**a**) a laser beam welded aluminium alloy (weld softened) and (**b**) a laser beam welded steel (weld hardened), according to Dos Santos et al. (2000) and Çam et al. (1999), see also Zerbst et al. (2014a)

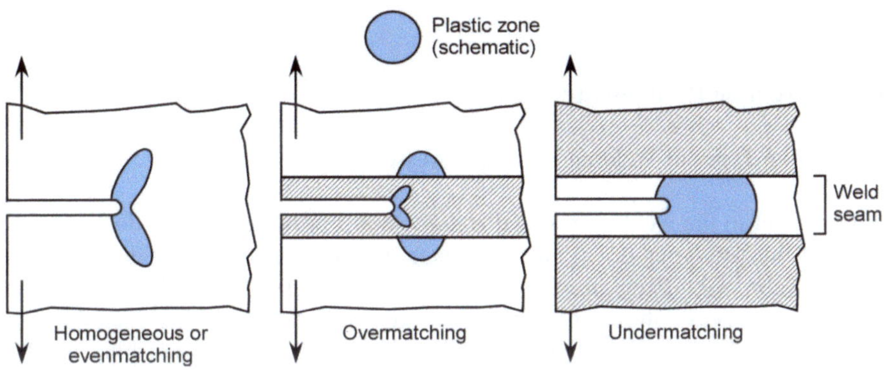

Fig. 32 Pattern of plastic deformation at the crack tip when affected by strength mis-match; schematic view; according to Annex P of BS 7910 (2013)

the higher strain. If the configuration contains a crack, strength mis-match will affect the formation of the plastic zone ahead of the crack tip. Its pattern will individually depend on the location of the crack with respect to the different material zones (Fig. 32). E.g., a crack at the center of an overmatched weld will be shielded (such as intended). If it is, however, located at the fusion line, the crack driving force will be increased due to the strain concentration in the less strength material. The consequence is a detrimental effect on the integrity of the structure. The situation becomes even more complex when the transition between the two materials is at a stress concentrator such as the weld toe.

Strength mis-match has clearly an effect on the fracture resistance of the material and on the monotonic failure load of the weldment; guidance on this will be given in Annexes B and C. However, since strength mis-match is a problem restricted to elastic-plastic deformation, the question arises whether there will also be an effect on small-scale yielding-controlled high cycle fatigue. Note that there are two indications that this should be the case:

(a) The plasticity-induced crack opening effect (see Sect. 1.2.2) is caused by the plastic zone ahead of the crack which remains as stretched material zone in the wake of the crack when this propagates. It is easy to imagine that strength mis-match will affect the shape and dimension of this plastic zone.
(b) In notched geometries such as weldments, residual stress relaxation can occur under monotonic as well as cyclic loading due to local plastic deformation driven by the stress concentration at the notch root. In this case too, strength mis-match might have an effect in that it affects the plasticity pattern at the notch root and the fusion line, see also the discussion in Sect. 1.9.4.2.

Note that both mentioned effects are much too complex to be treated in a manageable way by an analytical approach. Nevertheless, they are mentioned here because they might explain some empirical effects which will be briefly addressed below.

1.9.4 Welding Residual Stresses

1.9.4.1 Two Types of Welding Residual Stresses

With respect to the welding residual stresses it has principally to be distinguished between the two types of (a) *medium-range* and (b) *long-range* residual stresses. Note that the original definition (Green and Knowles 1992) contains *short-range* residual stresses as a third type. Since these generally refer to rather deep through-thickness cracks (see also Budden and Sharples 2003) they do not play much a role in the present context. Therefore, the following discussion is limited to types (a) and (b).

(a) **Medium-Range Residual Stresses**

These can also be designated by near-weld self-equilibrating stresses and correspond to what was named *secondary* stresses in Sect. 1.5.3. They are formed by two mechanisms: (i) locally varying shrinkage during the cooling process in and adjacent to the weld and (ii) phase transformation (e.g., from fcc to bcc lattice in ferritic or bainitic steels) with accompanying volume extension due to the lower packing density. Hindered shrinkage or extension reappears as residual stresses.

(b) **Long-Range Residual Stresses**

These are also designated as reaction stresses and they are *primary* stresses in terms of the nomenclature of Sect. 1.5.3 Different to the *medium-range* stresses the forces and moments are in equilibrium across the overall structure but not across the section potentially carrying the crack. They are formed due to structural mis-match (e.g., when a structural member is fitted into an assembly) which induces restraint. Usually reaction stresses are released when a weld detail specimen is cut from the structure.

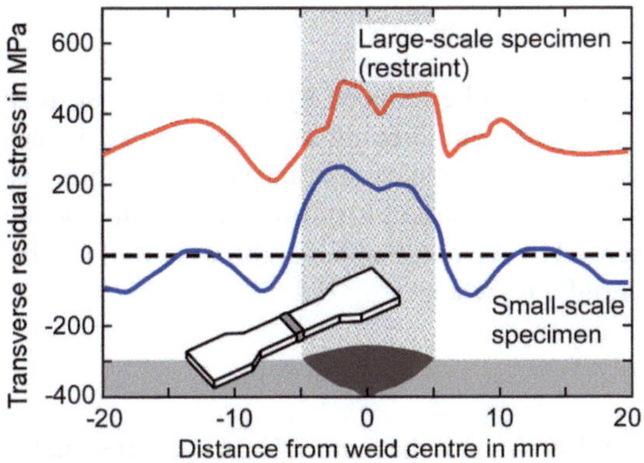

Fig. 33 Transverse residual stresses at the surface of butt welded plates with and without restraint; Steel S355; according to Farajian (2013)

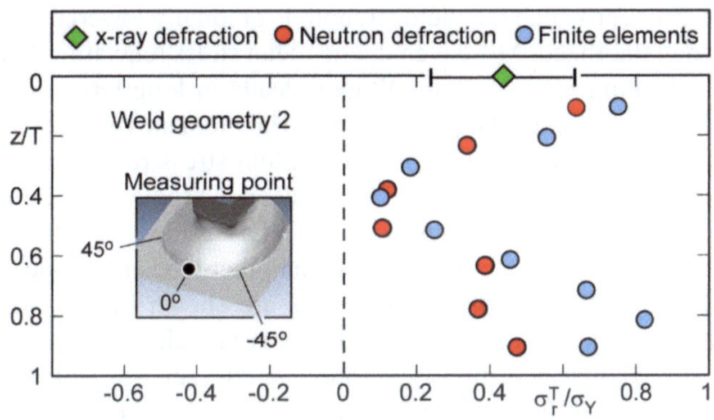

Fig. 34 Depth profile of residual stresses in a MAG-welded plate with longitudinal gusset made of S355NL investigated within IBESS; according to Hensel et al. (2018)

Note, however, that there might also be some kind of "internal restraint" in a weld detail specimen, e.g. in a longitudinal gusset. Reaction stresses are affected by a wide range of factors such as the geometry of the members to be joined, the use of jigs and other fabrication aids during welding, the pass sequence in multi-pass welds and the overall welding sequence when the structure comprises more than one weld (Leggatt 2008). An example of superimposed *medium-* and *long-range* stresses is provided in Fig. 33, an example for a residual stress profile in the thickness direction of a self-restrained structure in Fig. 34.

1.9.4.2 Redistribution and Relaxation of Residual Stresses

Welding residual stresses can be reduced by post-weld heat treatment (PWHT) or post-weld mechanical treatment. Peening methods modify the near-surface state of weldments by introducing compressive residual stresses. In any case, the basic mechanism of residual stress manipulation is plastic deformation which occurs when the applied stress (increased by the notch effect at the weld toe) locally reaches the yield strength of the material, i.e., when

$$\sigma^{p} \geq \sigma_{Y} - \sigma_{r} \tag{10}$$

with σ_Y being the yield strength of the material and σ_r the residual stresses. In PWHT, this effect is reached by lowering the yield strength σ_Y, in monotonic loading by a σ^p high enough to fulfil Eq. (10).

Cyclic loading (beyond the first loading cycle) will have an effect in cyclically softening materials for which the yield strength is gradually lowered until the stress-strain response is stabilized. Moreover, cyclic mean stress relaxation (which is more pronounced in loading with high amplitudes) leads in long term to a local stress strain path which approaches the path of a material element without residual stress. An example of experimentally determined residual stress relaxation due to cyclic loading is shown in Fig. 35, a numerical analysis on the topic was performed within the IBESS project on cruciform joints of S355NL. Besides the effect of the stress amplitude (as expected the effect increases with higher load amplitude) this analysis revealed an R ratio effect. Parts of the results are shown in Fig. 36. At $R = -1$ the relaxation of both initial tensile and compressive residual stresses was observed whilst at $R = 0$ a gradual built-up of compressive stresses was stated and that also for both initial states. In both the experimental as well as the numerical analysis, transverse residual stresses at the weld toes of butt welds of S355 steel grade were investigated. For a more detailed discussion see Hensel et al. (2018).

1.9.4.3 Treatment of Residual Stresses in Fracture Analysis

1.9.4.3.1 Determination of Residual Stress Profiles

The determination of (welding) residual stress profiles in components is anything but simple, in particular as the *long-range* or reaction stresses will vary from structure to structure. Usually this information has to be provided by finite element simulation or experimental determination on the structure. Documents such as BS 7910 (2013) provide reference profiles of residual stresses at the plate surface as well as in the thickness direction. However, these are often not really helpful for a couple of reasons:

(a) The compendium distributions are upper bound curves to the data of literature reviews. They are not only conservative but can even be misleading. Figures 37b provides surface distributions of transverse residual stresses (acting as mode-I stresses in fracture mechanics) across the weld for steels with yield strengths between 300 and

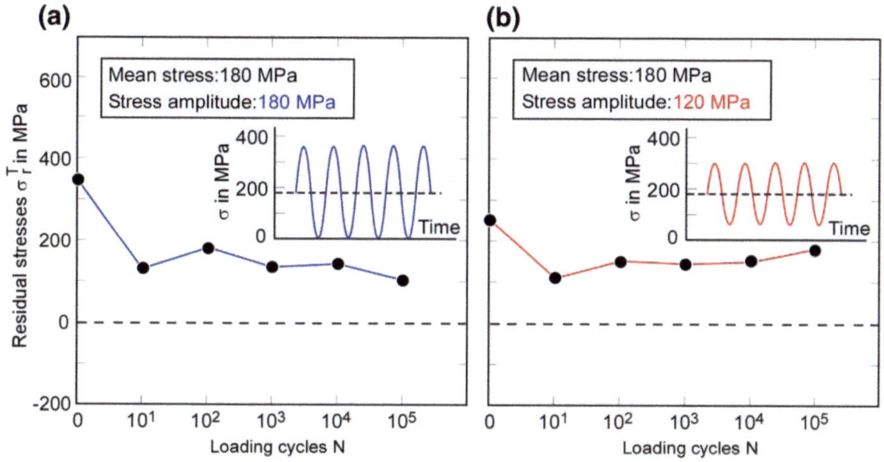

Fig. 35 Cyclic residual stress relaxation in the middle of the weld seam of a MAG butt-welded plate made from S355NL; according to Nitschke-Pagel (1995)

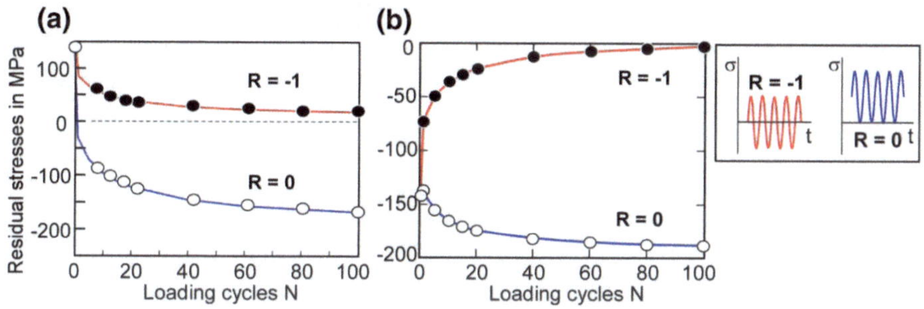

Fig. 36 Finite element simulation of the modification of as-welded residual stresses at the weld toe. **a** Initially tensile; **b** initially compressive residual stresses under cyclic loading at remote applied stress $\sigma_a = 100$ MPa; according to Hensel et al. (2018); see also Tchoffo Ngoula et al. (2018a)

1000 MPa (Fig. 37a) (Farajian 2013). As can be seen, there is a significant difference between the residual stresses at the centre of the weld and at the weld toe. As a rule, high tensile residual stresses are found at the first position whereas relatively low tensile or even compressive residual stresses are determined at the weld toe. This pattern was also found in IBESS. Note that all distributions refer to *medium-range* residual stresses, i.e., *long range* or reaction stresses were absent.

In Fig. 38 transverse residual stress depth profiles below the weld toe of S355NL steel are shown. The two experimental profiles which have been obtained in IBESS by combined x-ray and neutron diffraction and hole drilling and are supplemented by finite element simulation using the software SYSWELD® refer to MAG and TIG welding. For comparison, the corresponding compendium profile of BS 7910 (2013)

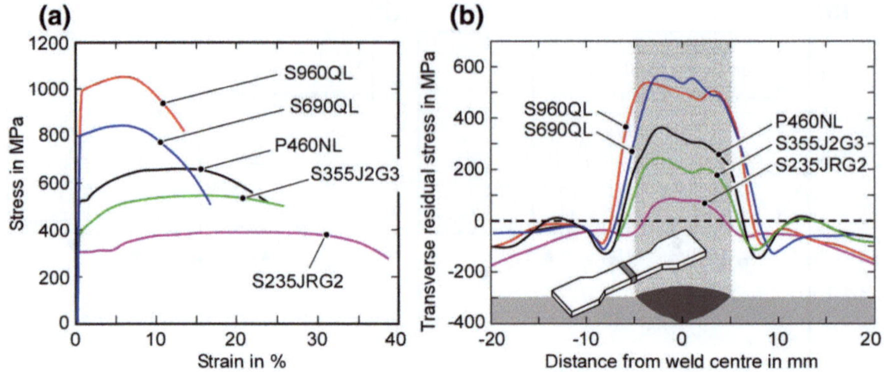

Fig. 37 Transverse residual stresses at the surface of butt welded plates. **a** Monotonic stress-strain curves of the materials investigated; **b** residual stress profiles at different positions including base plate, weld and weld toe; according to Farajian (2013)

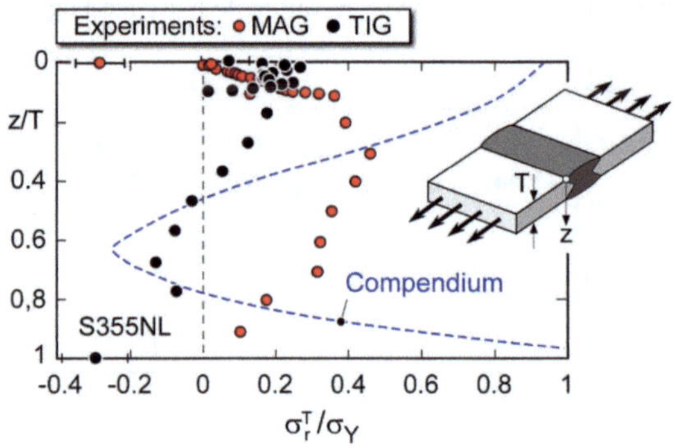

Fig. 38 Experimental transverse residual stress profiles at the weld toe of S355NL steel butt welds investigated within IBESS as compared with the upper bound solution of the BS 7910 (2013) compendium; according to Hensel et al. (2018)

$$\sigma_r^T / \sigma_Y = 0.9415 - 0.0319(z/T) - 8.3394(z/T)^2 + 8.660(z/T)^3 \tag{11}$$

is inserted. A big difference is seen. Note that for the fracture mechanics-based assessment of the fatigue strength the stress profile beneath the surface up to a depth of ½ or one millimeter is crucial which means that the compendium profile can simply not be used for the present task. The main cause of the difference is that Eq. (11), as an upper bound, covers the conditions at the center of the weld seam whereas the fatigue crack, in the present context, is initiated at the weld toe.

(b) At least part of the information collected in the compendia is taken from examples with rather undefined mixtures of *medium-* and *long-range* residual stresses, e.g., some profiles have been obtained on tubular joints (Bate et al. 1997). This might not only differ from the applications for which the profiles are to be used, it also brings up the question on whether they are *primary* or *secondary*. Note, however, that some sensitization had happened with regard to this issue and that recent updates of compendia (BS 7910, 2013; R6, 2014) try to avoid or at least to minimize the *long-range* residual stresses.

1.9.4.3.2 Residual Stress Classification

Provided the residual stresses are known, they have to be classified as *primary* or *secondary* in the next step. The general definition in fracture mechanics is that *primary* stresses contribute to plastic collapse but *secondary* stresses do not. As the consequence, they will be treated in different ways in elastic-plastic fracture mechanics assessment.

However, the classification must be based on criteria independent of its use. Sometimes, as shown in Fig. 33, the decision is easy, but there are many cases where the distinction is a sophisticated task or even impossible. In doubt, the classification as *primary* will be conservative. The situation can also be described by a phenomenon called "elastic follow-up" which characterizes the ability of a structure to increase local strains by plastic (or creep) deformation while under constant load or displacement (McClung et al. 1999).

The basic idea is illustrated by the three-bar model shown in Fig. 39. In order to fit the middle bar into the frame it has to be stretched whereas the two outer bars are compressed (Fig. 39a). Under elastic conditions (the outer bars act like springs on the middle bar, Fig. 39b) the stresses in the bars depend on (i) the initial misfit in the bar's lengths, (ii) the stiffness of the bars and (iii) the relative stiffness of the assembly. Now let us assume that the middle bar is not a single entity but contains a weld-like reinforcement (Fig. 39c). At this position the stiffness is different from the rest of the bar and there are stress concentrations at the transitions. Although the residual stresses acting at the middle bar are (elastic) *long-range* stresses, they can cause plastic deformation at the local notches above a certain stress level. If we think about a real weld, the near-weld self-equilibrating stresses can be affected this way. In other words: the *long-range* residual stresses will interfere with the *medium-range* stresses due to the elastic follow-up phenomenon. Note that the effect will not only depend on the magnitude of the reaction stresses and on the local geometry but also on potential stress relief due to plastic deformation and crack extension at the notches. These reflections provide us with an alternative definition of *primary* and *secondary* (residual) stresses in that *primary* residual stresses exhibit significant elastic follow-up whilst *secondary* residual stresses do not.

Based on this definition, residual stresses exhibiting significant elastic follow-up will generally be classified as *primary*. This is usually the case when they arise from *long-range* restraint effects. If they are caused by self-restraint of the welding detail (such as with longitudinal gussets) the situation may be more difficult (Budden amd Sharples 2003) but classification as primary is at least at the safe side. With respect to *short* cracks which are particularly in the focus in the present context since they largely contribute to the total lifetime of the structure, IBESS follows the proposal of BS 7910 (2013) to classify residual stresses generally as *primary* when their spatial extent is large compared with the defect size. Note, however, since the effects of elastic follow-up are reduced in the case of widespread plasticity this is not recommended for low cycle fatigue (LCF).

Note further that the concept of elastic follow-up is also related to the proximity of the weld to a load- or displacement-controlled situation in that it is high for the

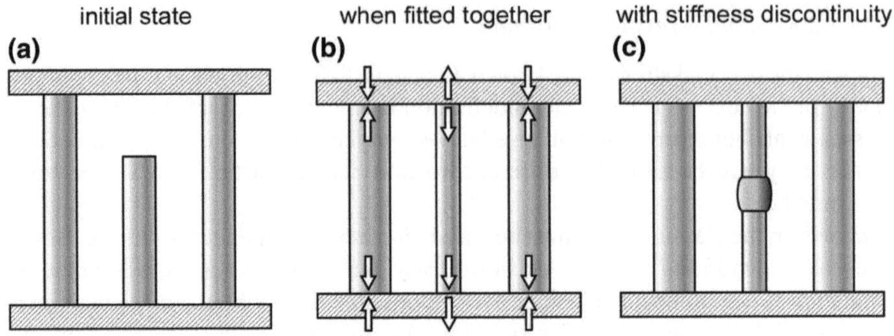

Fig. 39 Three-bar model for explaining the elastic follow-up phenomenon

first and insignificant for the latter one (BS 7910, 2013). A load-controlled situation is expected when the locality of the weld is far from the displacements whereas a displacement-controlled situation refers to a locality close to the displacements.

1.9.4.3.3 The Interaction Factor V

If residual stresses are classified as *primary*, things are straightforward in that they are treated like any other applied load. If they are classified as *secondary*, they are only considered in stress intensity factor determination but do not influence the ligament yielding parameter L_r, see Annex C. For small scale yielding conditions the total stress intensity factor for mode-I loading, K_I, can simply be determined by superposing the K_I factors K_I^p and K_I^s for *primary* and *secondary* loading:

$$K_I = K_I^p + K_I^s \tag{12}$$

However, in the general case, plasticity and relaxation effects must be considered. Although they do not contribute to plastic collapse, *secondary* stresses (such as the *primary* ones) cause ligament yielding with the consequence that the resulting crack driving force exceeds $K_I^p + K_I^s$ in the contained yielding regime. On the other hand, the total K_I is reduced to a value below $K_I^p + K_I^s$ for net section yielding due to residual stress relaxation. Both effects are considered by a factor V multiplied to K_I^s such that K_I becomes

$$K_I = K_I^p + V \cdot K_I^s \tag{13}$$

The dependency of V on the degree of ligament yielding, expressed by L_r, is schematically illustrated in Fig. 40. Besides this, V depends on the magnitudes of the *primary* and *secondary* stresses and on the crack size.

V factors have been obtained by finite element analyses and they are provided in table format, e.g., in Sect. 2.6 of R6 (2014) or as parametric equations such as in Annex R of BS 7910 (2013). No detailed discussion will be provided here on the common approach since this is part of the two documents mentioned above. Considerable effort was spent during the last decade to reduce the conservatism in fracture analyses with residual stresses. Two new items are

(a) the use of the same ligament yielding correction f (L_r) function for the plasticity correction of K_I^s which is also applied to the *primary* stresses, and
(b) the incorporation of elastic follow-up by a factor Z which is usually obtained by finite element analyses. Advice on this is given in the R5 document (2014).

Note that the advanced procedures provide alternative, less conservative options to the treatment of the residual stresses as *primary* even in the case of elastic follow-up in that they are treated as *secondary* but in conjunction with an increased value of V.

Note further, that the relevance of these improvements in the present context, i.e., for fatigue strength determination is rather limited since the total lifetime is much more controlled by the *short* crack propagation stage than by the criteria for final fracture.

1.9.4.4 Treatment of Residual Stresses in Fatigue Crack Propagation

The discussion within this section is restricted to analytical analyses based on weight function or similar solutions for the stress intensity factor, see Sect. 2.2.3.2.5. The advantage of this kind of solution is that it is based on the stresses in the uncracked body and no stress redistribution due to crack propagation has to be taken into account. Within the linear elastic ΔK concept and in the absence of crack closure effects, the principle how to consider residual stresses is straightforward. Whilst the stress intensity factor range ΔK itself is not affected by the residual stress-induced stress intensity factor K_r, the latter modifies the mean stress $\bar{K}$

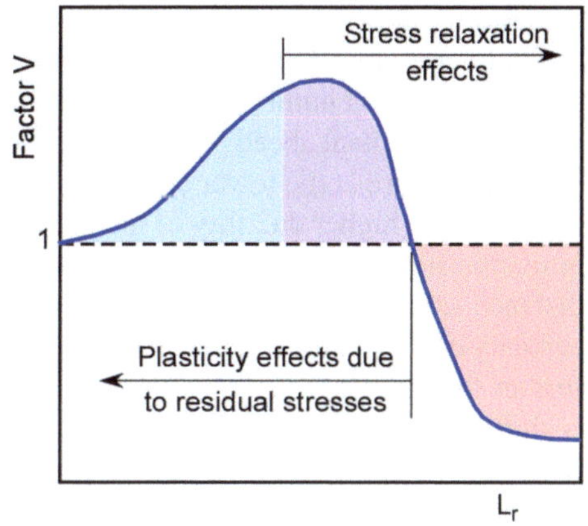

Fig. 40 The V correction term as a function of the ligament yielding parameter L_r (schematic)

$$\bar{K} = \frac{1}{2}[(K_{max} + K_r)+(K_{min} + K_r)]=\frac{1}{2}[K_{max} + K_{min}] + K_r \tag{14}$$

and the R ratio

$$R = \frac{K_{min} + K_r}{K_{max} + K_r} \tag{15}$$

which is then understood as an effective rather than the nominal R ratio exclusively based on the applied loading. The crack propagation analysis in the structure uses the da/dN–ΔK curve obtained for this effective R ratio.

However, when crack closure comes into play, simple linear superposition is not adequate any more and both, the R ratio and ΔK_{eff} are affected by the residual stresses. In this case an additional correction for crack closure has to be performed as it is, e.g., done by Newman's approach which is also part of the so-called NASGRO curve (Newman 1984; NASGRO 2000), see Sect. 2.2.4.2.

Note that BS 7910 (2013) offers a simplified solution for weldments in that it always assumes high tensile residual stresses which increase R, this way keeping the crack open. Consequently the document offers reference da/dN–ΔK curves for $R \geq 0.5$ for use in conjunction with weldments, see Sect. 2.3.2.5.3. However, there might also be compressive residual stresses which lower the crack opening load, see, e.g. Beghini et al. (1994) and Sect. 1.9.4.2. In such a case the fatigue crack propagation analysis would be conservative, i.e. the crack growth rate would be overestimated.

1.9.5 Specific Aspects Concerning the Testing of Welds

1.9.5.1 Microstructural Inhomogeneity and Fracture Resistance

Microstructural inhomogeneity causes a scatter in fracture resistance, in particular if the failure follows a weakest link mechanism. This is, e.g., the case in the ductile-to-brittle fracture regime of materials with bcc or hexagonal crystal lattice, see Figs. 29 and 30. A limited number of microstructural flaws is stochastically spread over the uncracked ligament ahead of the crack. Note that most of the flaws which would have triggered failure at the lower shelf are "defused", i.e. the micro-notches are rounded off due to the higher ductility of the material. A limited number of defects, however, remains critical. When the structure is loaded, a stress peak forms at the ligament at a distance 1.6–2.4 times the crack tip displacement (depending on the strain hardening capacity of the material, Chen and Cao 2015) which is shifted into the ligament with further load increase and crack extension. Failure is triggered when the stress peak is shifted to the position of the one flaw the position of which is closest to the crack front.

Due to the stochastic distribution of the flaws across the ligament the position of this flaw differs from specimen to specimen. Thus, the amount the stress peak has to

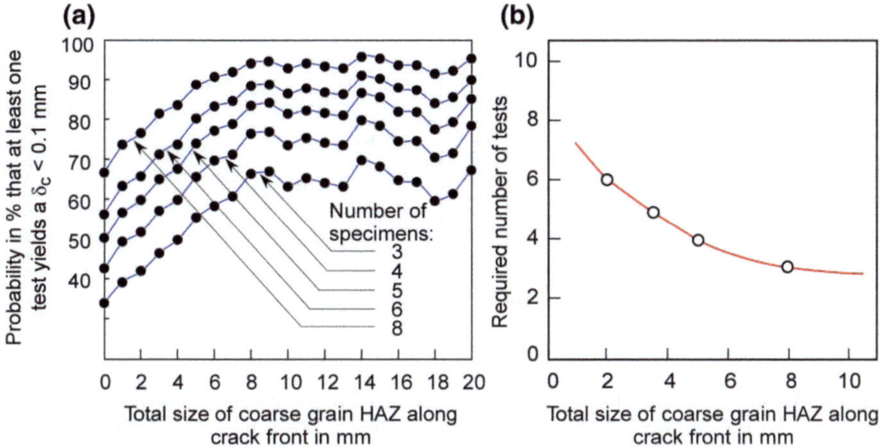

Fig. 41 **a** Probability that at least one low toughness value is obtained in a test set as a function of the total length of the coarse-grained heat affected zone (CGHAZ) cut by the crack front and the number of specimens tested; **b** required number of fracture toughness tests versus the total length of the coarse-grained heat affected zone cut by the crack front; according to Toyoda (1989)

be shifted into the ligament up to fracture is also different and so is the energy to be spent. This way, a significant scatter band of the fracture resistance is obtained. Note that the probability of a flaw close to the crack front increases with the crack front length such that the scatter band is "squeezed" towards its lower bound (statistical geometry effect on monotonic fracture resistance).

In welds, local brittle zones, usually of coarse grain HAZ microstructure, may act as microstructural flaws. As a consequence, the width of the scatter band reduces and the probability to determine a lower bound fracture resistance value increases with the proportion of the crack front located in brittle microstructure areas. This is the background of requirements of test standards such as BS 7448, Part 2 (1997) and ISO 15653 (2010) for pre- and post-test metallography for generating and validating a sufficiently large percentage of brittle zone material sampled by the crack front and of a sufficiently large number of tests.

Note that from the viewpoint of weakest link statistics an insufficient volume of HAZ microstructure in one specimen can be compensated by testing a larger number of specimens. The principle is illustrated in Fig. 41 (Toyoda 1989). Figure 41a provides the probability that, in a test set, at least one low toughness value is obtained as a function of both the total length of the coarse grain heat affected zone sampled by the crack front and the number of specimens tested. This information is further condensed in Fig. 41b where both parameters are plotted against each other.

In that context, the 2005 version of BS 7910 required a minimum of 12 valid test results to characterize HAZ toughness distributions in structural steels with yield strengths up to 450 MPa, whilst the 2013 update speaks about "a larger number

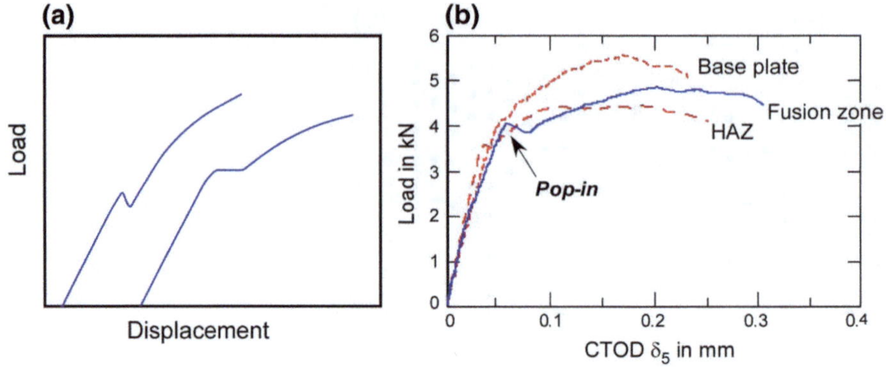

Fig. 42 Pop-in events in the load-displacement record. **a** Types according to Dawes et al. (1989); **b** example: Laser beam weld of a titanium alloy; according to Dos Santos et al. (2000)

of test results … the number" of which being "dependent on the type of statistical analysis". In the upper shelf, three specimens are regarded as sufficient.

In principle, HAZ fracture resistance data can be determined in two different ways:

(a) By testing weldment specimens prepared such that the front of the pre-crack is located within the desired weld region and post test evaluation to test the efficiency of this preparation. The methodology is described in test standards BS 7448, Part 2 (1997) and ISO 15653 (2010), see also Zerbst et al. (2014a).
(b) By testing specimens consisting of thermally simulated microstructure.

When the HAZ microstructure is satisfactorily reproduced, option (b) offers various advantages, i.e., a possible reduction of the number of specimens and the avoidance of crack path deviation away from the brittle zones, e.g., in strength mis-matched configurations.

A last point to be mentioned here is the susceptibility of many welds to pop-in behaviour (Dawes et al. 1989). A pop-in is a discontinuity in the load-displacement record in a fracture mechanics test such that the displacement temporarily increases and/or the load decreases (Fig. 42). In weldments this phenomenon is usually caused by crack initiation at local brittle zones followed by crack arrest in the surrounding more ductile material but there might be other reasons as well, e.g., the formation of splits or delaminations perpendicular to the fatigue crack plane, coalescence between multiple cracks or cracks with other flaws (slag inclusions, pores), etc. The treatment of pop-ins in monotonic fracture testing is also briefly addressed by Zerbst et al. (2014a). Note that pop-in events may also occur in fatigue crack propagation where they increase the slope n of the da/dN–ΔK curve. They are quite typical in *Region* 3 of the diagram, see Sect. 1.4.1, Fig. 5.

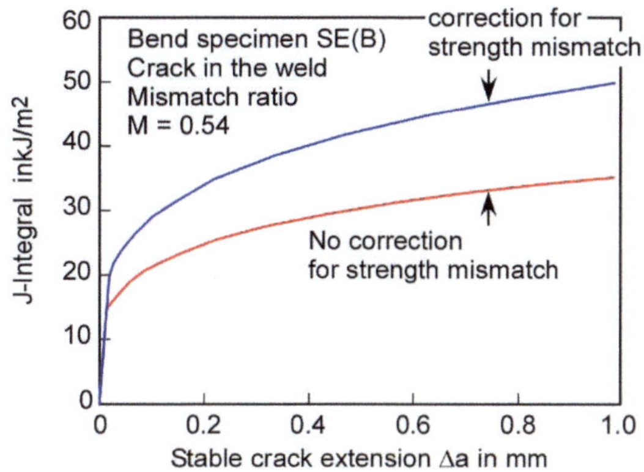

Fig. 43 Example for strength mis-match corrected R-curves (Zerbst et al. 2014a), see Annex B

1.9.5.2 Strength Mis-match

Documents such as EFAM GTP (Schwalbe et al. 2002) and ISO 15653 (2010) provide limits of the mis-match ratio M within which strength mis-match effects do not have to be considered. Usually the accepted error in the J-integral is in the order of 10%. Beyond these limits, the EFAM GTP document (Schwalbe et al. 2002) offers an option for mis-match correction of over- and undermatched SE(B) and M(T) specimens in which the common J calibration function η_p in

$$J = \frac{K^2}{E} + \eta_p \frac{U}{B(W - a)} \tag{16}$$

is replaced by a mis-match corrected one, see Annex B. Figure 43 provides an example of a crack resistance curve $(J - \Delta a)$ corrected this way.

Strength mis-match has also an effect on the stress triaxiality, which controls the constraint in the ligament of a cracked specimen (Schwalbe et al. 1997). An example is provided in Fig. 44. Note that a higher constraint not only reduces the crack resistance but may also affect the fracture mechanism. As the result, the failure regime of materials with transition behaviour may change from the upper shelf to the ductile-to-brittle transition regime, cf. Sect. 1.9.2.3. Another effect is crack path deviation such that the crack tip during crack extension samples quite different material conditions.

1.9.5.3 Welding Residual Stresses

Welding residual stresses are treated on the crack driving force side in a fracture mechanics analysis which means that they should be avoided on the material test side. This makes sense since specimen preparation from a larger configuration already causes some residual stress relief. Nevertheless, in addition, some special specimen preparation is required by the test standards (BS 7448, Part 2, 1997; Schwalbe et al.

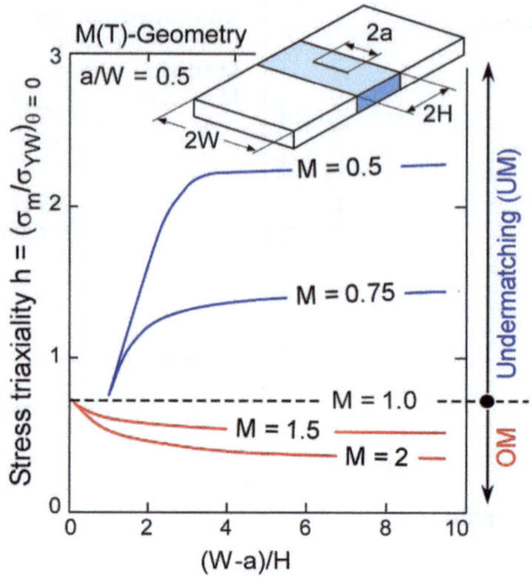

Fig. 44 Effect of strength mis-match on stress triaxiality at the crack tips of a center crack in a welded plate, example according to Schwalbe et al. (1997)

2002, [EFAM GTP]; ISO 15653, 2010) not only to remove (or at least homogenize) the remaining residual stresses but also to provide baseline conditions for valid tests, e.g., a largely straight crack front. The options for this are summarized in Fig. 45, for a discussion see Zerbst et al. (2014a).

Note that these measures, in principle, allow the determination of monotonic crack resistance but they fail with respect to fatigue crack propagation because potential crack path deviation makes it almost impossible to provide the material characteristic of a specific material zone such as the HAZ. In many cases, the only alternative is to use specimens with thermally simulated microstructures such as briefly addressed above.

2 The IBESS Approach

2.1 Basic Philosophy

Comparable to existing procedures, e.g. R6 (2014), BS 7910 (2013) or SINTAP (1999), see also Webster and Bannister (2000) and Zerbst et al. (2007) the IBESS draft document is organized along various analyses levels which differ in the quality of the input data, the complexity of the analysis and the conservatism of the results. Note that the structure of the procedure presented here is tentative and that comprehensive validation is still lacking. In its current state it will undoubtedly need improvement. In some cases, significant effort must be driven at the higher analysis levels which is difficult to achieve under normal conditions. For the latter, simplified estimation methods at a lower analysis level are provided. The following sections will mainly describe what the authors call the IBESS basic procedure whilst proposals for lower analysis levels will be added by inserted subsections, e.g., 2.3.1.7, 2.3.2.2b, 2.3.2.3b, 2.3.2.4.3, 2.3.2.5.3, 2.4.7 and 2.7.4.

2.2 *Cyclic Elastic-Plastic Crack Driving Force*

2.2.1 Monotonic J-Integral—Basic Definition

As mentioned in Sect. 1.1.3, the crack driving force of *mechanically short* cracks cannot be based on linear elastic fracture mechanics, i.e., the ΔK parameter common for *long* fatigue crack propagation has to be replaced by an elastic-plastic counterpart. In IBESS, this is the cyclic J-integral which is determined by finite elements and, preferentially, by analytical equations. Within this section, some background information on ΔJ will be provided.

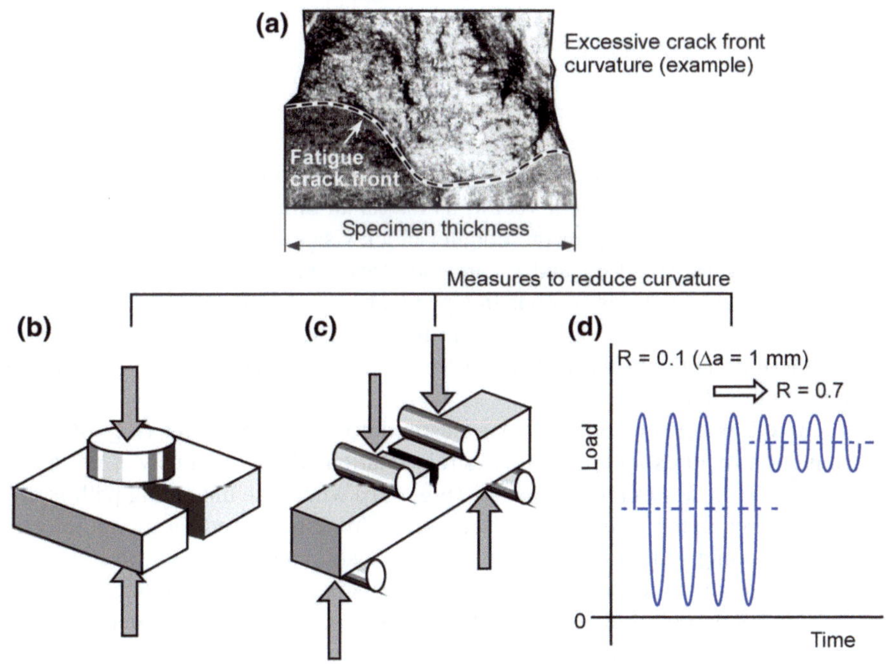

Fig. 45 Methods proposed for reducing and redistributing residual stresses in test specimens prior to and during pre-cracking; **a** example for excessive crack front curvature due to welding residual stresses (Figure according to Koçak et al. 1990); **b** local compression method; **c** reverse bending method; **d** stepwide high R ratio method; for details see Zerbst et al. (2014a)

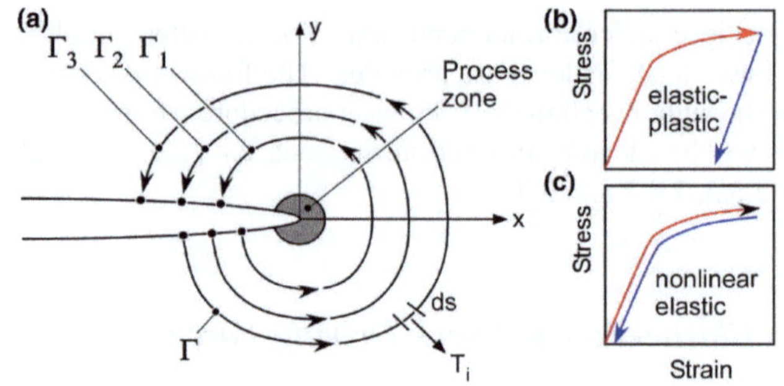

Fig. 46 **a** Contours for the determination of the J-integral, schematic view; **b** elastic-plastic; **c** nonlinear elastic material behaviour

The basic definition of the (monotonic) J-integral is given by

$$J = \int_{\Gamma} \left(W dy - T_i \frac{\partial u_i}{\partial x} ds \right) \tag{17}$$

Rice (1968). In Eq. (17), W is the strain energy density

$$W = \int_{0}^{\varepsilon_{ij}} \sigma_{ij} d\bar{\varepsilon}_{ij}, \tag{18}$$

T_i are the components of the traction vector normal to the contour Γ, u_i are the components of the displacement vector, ds marks an infinitesimal arc length along Γ surrounding the crack tip in the counter-clockwise direction from the lower to the upper crack face and σ_{ij} and ε_{ij} are the stress and strain tensor components (Fig. 46a). The parameter is contour-independent, i.e., identical values are obtained by integrating along different contours (in the Fig. Γ_1, Γ_2 and Γ_3 as examples) as long as these do not pass through the process zone immediately at the crack tip or touch the outer edge of the body.

Note that J is defined for nonlinear elastic deformation behavior (Fig. 46c). Its determination requires a unique stress-strain relation which obviously is not given for elastic-plastic deformation behavior (Fig. 46b) where the unloading path is different compared to the loading path. Because of this, J is defined for monotonically increasing loading only where the non-linear elastic and the elastic-plastic curve branches are identical. Unloading as it, e.g., occurs in the crack wake of a propagating crack is not permissible. When the J-integral, nevertheless, is used in monotonic R-curve testing, stable crack extension is restricted to small amounts to ensure that the unloading is confined to a small region at the crack tip.

2.2.2 Cyclic J-Integral ΔJ

2.2.2.1 Numerical Determination

The cyclic J-integral follows the definition of the monotonic J-integral, however, includes some specific requirements:

$$\Delta J = \int_{\Gamma} \left(\Delta W dy - \Delta T_i \frac{\partial(\Delta u_i)}{\Delta x} ds \right) \tag{19}$$

with

$$\Delta W = \int_{0}^{\Delta\varepsilon_{ij}} \Delta\sigma_{ij} d\left(\Delta\bar{\varepsilon}_{ij}\right) \tag{20}$$

Lamba (1975), Dowling and Begley (1976). The Δ symbol preceding the stress and strain tensors as well as the traction and displacement vector components designates the changes of these quantities

$$\Delta\sigma_{ij} = \sigma_{ij}^{max} - \sigma_{ij}^{min}, \tag{21}$$

$$\Delta\varepsilon_{ij} = \varepsilon_{ij}^{max} - \varepsilon_{ij}^{min}, \tag{22}$$

$$\Delta T_i = T_i^{max} - T_i^{min} \tag{23}$$

and

$$\Delta u_i = u_i^{max} - u_i^{min}, \tag{24}$$

however, in ΔJ and ΔW it does not represent changes of J and W; instead ΔJ and ΔW are functions of their arguments as defined by Eqs. (19) and (20). In other words: Different to the K-factor range there is no J-integral range such that $\Delta J \neq J_{max} - J_{min}$. Instead the term cyclic J-integral is common.

Note that for linear elastic deformation behavior things are easier in that an elastic ΔJ, ΔJ_e, can simply be correlated to ΔK by

$$\Delta J_e = (\Delta K)^2 / E' \tag{25}$$

with E' being E for plane stress and $E/\left(1-\nu^2\right)$ for plane strain conditions.

The ΔJ parameter defined in the proper way has successfully been applied as a parameter for the cyclic crack driving force. Nevertheless, it has also been faced with objections. The most serious one referred to the exclusion of crack growth in the basic definition of the J-integral (Eq. 17). How can ΔJ, as an extension of this parameter, be used for fatigue crack propagation where the loading path shows a hysteresis loop

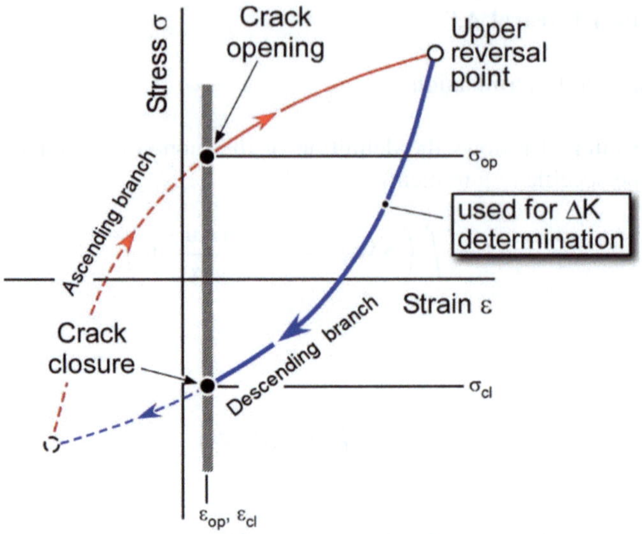

Fig. 47 Stress-strain hysteresis and the determination of ΔJ in the presence of the crack closure phenomenon, schematic

(Fig. 47) which means that the requirement for one unique stress-strain relation is definitely violated?

The argument is certainly correct, however, the problem is avoided if only the ascending (or the descending) branch of the stress-strain hysteresis with the reversal points as the reference states is used for the determination of ΔJ, for a more detailed discussion see Tchoffo Ngoula et al. (2018a, b). Both options are equally legitimate and yield identical results as long as the hysteresis loop exhibits symmetric behavior in tension and compression.

A second objection refers to crack closure. Note that the path-independency of J is violated when there are stresses at the crack faces which is the case when these are in contact. Note further that material points at various locations near the crack tip and in the far field correspond to different positions at the ascending hysteresis branch at the moment of crack opening (Vormwald and Seeger 1991). Because of this, the treatment of the crack closure phenomenon in ΔJ is not practicable when based on the lower reversal point, but the upper reversal point offers a possibility when, instead of the ascending, the descending branch down to crack closure is used (Vormwald and Seeger 1991; Vormwald 2014), see also Tchoffo Ngoula et al. (2018a, b).

Finally, it has to be mentioned that the path independency of ΔJ is not given any more if a material is not completely cyclically stabilized because no single equation describes the stress-strain behavior in the transient regime of cyclic hardening or softening. For similar reasons, path-independency should not be expected in the presence of temperature gradients (Saxena 1998).

2.2.2.2 Analytical Determination

With respect to approaches for the analytical determination of ΔJ an overview has been provided by Zerbst and Madia (2018a). Its determination in IBESS follows a similar philosophy as outlined above in that the Δ symbols do not wrongly designate a change in J but refer to changes in the arguments which are the K-factor and the applied stress σ_{app} in that case.

$$\Delta J = \frac{\Delta K^2}{E'} \cdot [f(\Delta L_r)]^{-2} \quad (26)$$

Zerbst et al. (2012b), Madia et al. (2017b). This expression is based on the reference stress approach of Ainsworth (1984). Note, however, that deviating from the latter, the parameter $\Delta\sigma_{app}$ is given as the applied cyclic stress (referring to the gross cross-section) and σ_0 is the reference yield stress which has been introduced to improve the accuracy of the estimate (Madia et al. 2014), see Sect. 2.2.3.2.6. With respect to basic aspects of this parameter see also Zerbst et al. (2012a). The function $f(\Delta L_r)$ is a cyclic plasticity correction function based on a cyclic ligament yielding parameter ΔL_r defined as

$$\Delta L_r = \frac{\Delta\sigma_{app}}{2 \cdot \sigma_0}. \quad (27)$$

The $f(\Delta L_r)$ function is taken from its monotonic counterpart in R6 (2014), BS 7910 (2013) and SINTAP (cf. Zerbst et al. 2007). Within IBESS, the highest analytical analysis option is applied:

$$f(\Delta L_r) = \left[\frac{E \cdot \Delta\varepsilon_{ref}}{\Delta\sigma_{ref}} + \frac{1}{2}\frac{\Delta L_r^2}{E \cdot \Delta\varepsilon_{ref} / \Delta\sigma_{ref}}\right]^{-1/2}. \quad (28)$$

The stresses σ_{ref}, respective $\Delta\sigma_{ref}$, and strains ε_{ref}, respective $\Delta\varepsilon_{ref}$, refer to the stabilized cyclic stress-strain curve, see Sect. 2.3.2.3.

The factor 2 in the denominator of Eq. (27) is formulated for materials showing a symmetric hysteresis in tension and compression. Note that the ΔJ determined by Eq. (26) is formally transferred into a "plasticity-corrected" ΔK, designated by ΔK_p

$$\Delta K_p = \sqrt{\Delta J \cdot E'} \quad (29)$$

for its use in fatigue crack propagation analyses. Figure 48 shows an example for the comparison of numerically and analytically determined ΔJ values, for more detailed information see Tchoffo Ngoula et al. (2018b) and Madia et al. (2017b).

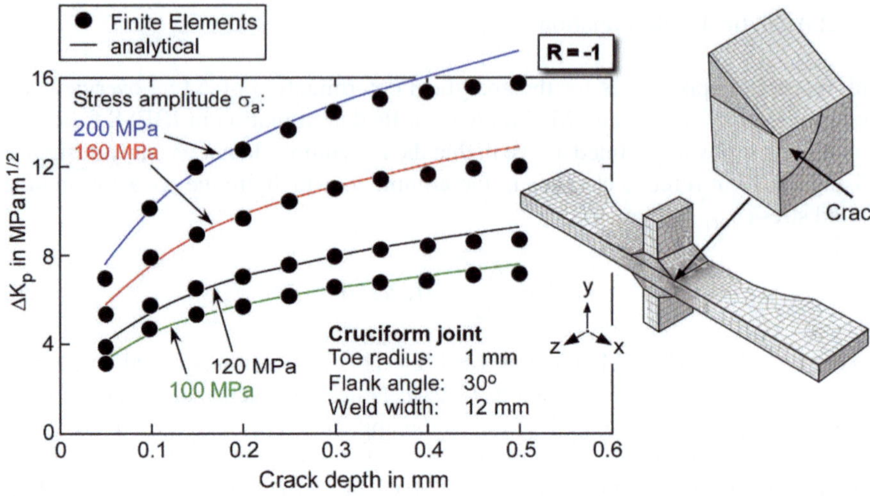

Fig. 48 Comparison between analytical and finite element based $\Delta K_p = \sqrt{\Delta J \cdot E'}$ for a cruciform joint of steel S355NL investigated within IBESS

2.2.3 IBESS Basic Procedure for ΔJ and ΔK_p Determination

2.2.3.1. General Scheme

A scheme for the determination of ΔJ and ΔK_p within the IBESS basic procedure is provided in Fig. 49. It comprises different steps which will be outlined in the following. Note that *Steps 2–3* can be replaced by two-dimensional finite element calculations at an advanced analysis level. An example is provided in Fig. 50. Likewise, the K-factor (*Step 5*) can be obtained numerically. At the highest analysis level, ΔJ is determined by finite elements based on Eq. (19)ff. Guidance on this is given by Tchoffo Ngoula et al. (2018b).

2.2.3.2 Analysis Steps

2.2.3.2.1 Selection of the Geometric Parameters of the Weld Toe

These parameters comprise the weld toe radius ρ, the flank angle α, the weld reinforcement h, and the secondary notch depth k, see Fig. 17 in Sect. 1.5.3. No guidance will be given at this point on the determination of these parameters, see, however, Sect. 2.3.1. Note that the parameters will be provided as statistical distributions for the IBESS basic procedure and as lower bounds, e.g., based on an extended weld quality catalogue at a lower analysis level, see Sect. 2.3.1.7.

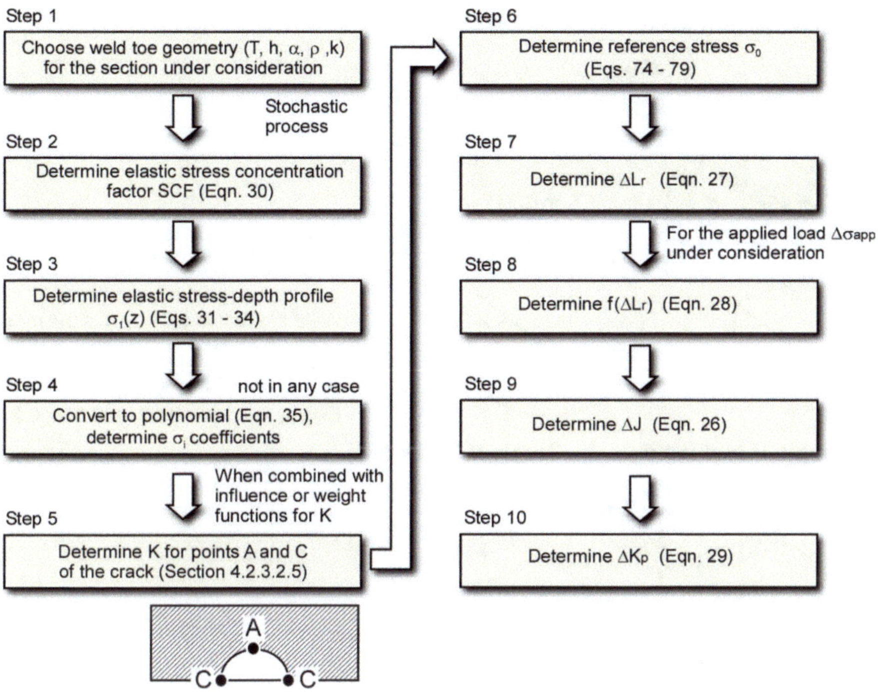

Fig. 49 Working steps for the analytical determination of ΔJ and ΔK_p in IBESS

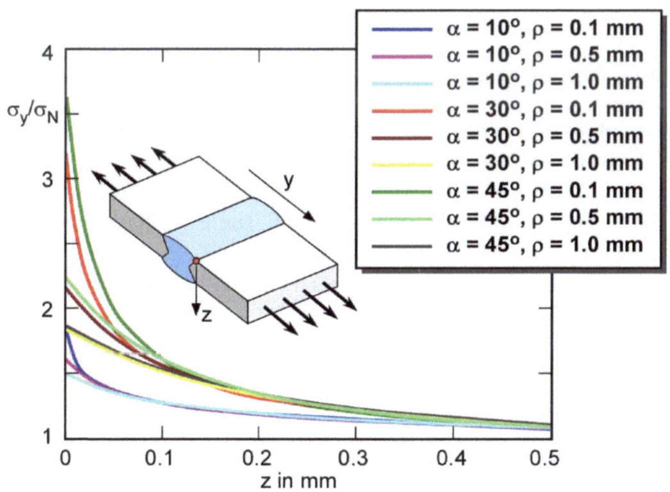

Fig. 50 Examples of finite element based through-thickness stress profiles (normalized) for some of the weldments investigated in IBESS

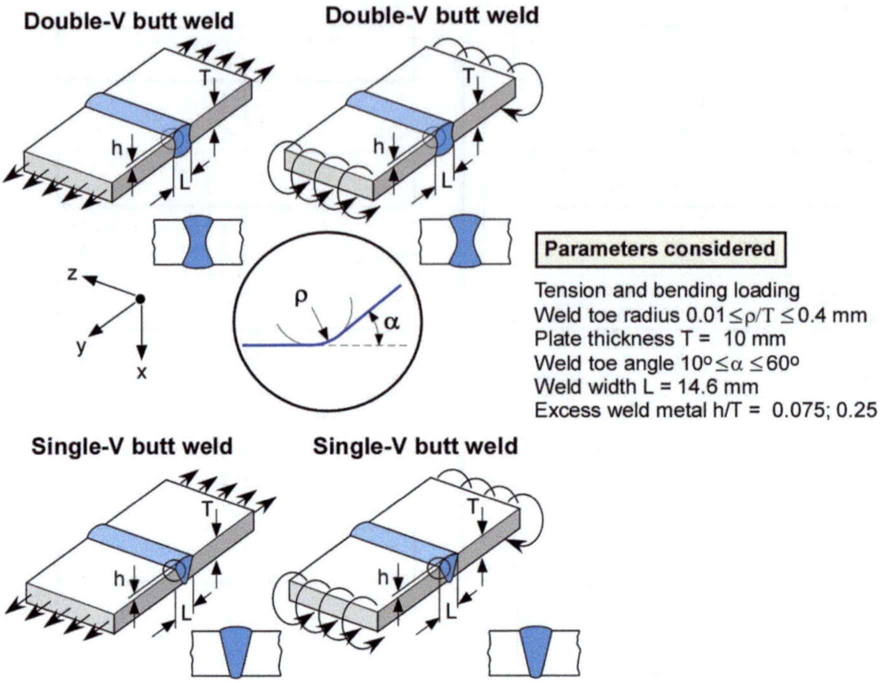

Fig. 51 Application range of the IBESS stress concentration factor (SCF) through-wall stress profile solutions (Eqs. 30 and 31ff)

2.2.3.2.2 Determination of the Elastic Stress Concentration Factor

This is needed for the analytical determination of the through thickness stress profiles in *Step 3*. Various parametric equations exist in the literature but a new variant was developed within IBESS which covers a wider range of geometry, particularly with respect to the weld toe radius ρ and the flank angle α (Kiyak et al. 2016). The application range is given in Fig. 51.

The IBESS solution is provided by Eq. (30) which is a modification of the solution of Brennan et al. (2000):

$$\mathrm{SCF} = 1 + p_1 \cdot \left(\frac{h}{T}\right)^{p_2 \cdot \alpha} \cdot \alpha^{p_3} \cdot e^{-p_4 \cdot \alpha} \cdot \left(\frac{\rho}{T}\right)^{-0.295 \cdot \alpha} \cdot \left(0.021 + \frac{\rho}{T}\right)^{-p_5} \tag{30}$$

with the coefficients p_1–p_5 being given in Table 1. The angle α is used in radians. For a weld reinforcement h considerably higher than 2.5 mm (e.g., T or cruciform joints) a maximum value of h = 2.8 mm is chosen. This is assumed to give a rather conservative value of the SCF.

Table 1 Coefficients of Eq. (30)

Configuration	p_1	p_2	p_3	p_4	p_5
Double-V, tension	1.9220	0.3224	1.1257	1.5481	0.4002
Double-V, bending	1.1399	0.2062	1.0670	1.6775	0.4711
Single-V, tension	1.3905	0.2081	1.0756	1.7483	0.4413
Single-V, bending	1.5326	0.2857	1.1036	1.5436	0.4287

2.2.3.2.3 Determination of the Through-Thickness Stress Profile

As with respect to the stress concentration factor of *Step 2*, parametric equations for the stress-depth profile have been obtained in IBESS (Kiyak et al. 2016) as an extension of an earlier equation of Monahan (1995). The application range is the same as in Fig. 51. The stress refers to the first principal stress σ_1 which acts as mode-I stress in the present configurations.

$$\sigma_I(z) = \begin{cases} \frac{SCF \cdot \sigma_n}{2\sqrt{2}} \cdot \left[\left(\frac{z}{\rho} + \frac{1}{2}\right)^{-1/2} + \frac{1}{2}\left(\frac{z}{\rho} + \frac{1}{2}\right)^{-3/2}\right] \cdot \frac{1}{G_l} & \text{for tension} \\ \frac{SCF \cdot \sigma_n}{2\sqrt{2}} \cdot \left[\left(\frac{z}{\rho} + \frac{1}{2}\right)^{-1/2} + \frac{1}{2}\left(\frac{z}{\rho} + \frac{1}{2}\right)^{-3/2}\right] \cdot \frac{1}{G_l} \cdot \left(1 - 2\frac{z}{T}\right) & \text{for bending} \end{cases} \tag{31}$$

with

$$G_\ell = \begin{cases} 1 & \text{if } z/\rho \le (z/\rho)_0 \\ q_1 + \frac{q_2 \cdot e^{-E_l \cdot T_l}}{1 + E_l^{q_3} \cdot T_l^{q_4} \cdot e^{-E_l \cdot T_l}} & \text{if } z/\rho > (z/\rho)_0 \end{cases}, \tag{32}$$

$$T_\ell = \frac{z}{T} - \left(\frac{z}{\rho}\right)_0 \cdot \frac{\rho}{T}, \tag{33}$$

and

$$E_\ell = q_5 \cdot \left(\frac{\rho}{T}\right)^{q_6 \cdot \alpha^{q_7}} + 1.5\alpha^{q_8 \cdot \rho^{q_9}} - q_{10} \cdot \left(\frac{h}{T}\right)^{q_{11} \cdot \alpha} - q_{12} \cdot \frac{h \cdot \alpha}{\rho}. \tag{34}$$

The coefficients q_1–q_{12} are given in Table 2.

Note that the same equations are used for single-V butt welds and T joints on the one hand and for double-V butt welds and cruciform joints on the other hand. Justification for this is provided in Fig. 52. A major advantage of using parametric equations for through-wall stress profiles is that this, compared to finite element analyses, provides a convenient and numerically efficient option for the probabilistic assessment when the weld toe geometry varies along its length dimension.

Table 2 Coefficients of Eqs. (31–34)

Weld geometry	Double-V	Single and double-V	Single-V
Loading (ℓ)	Tension (m)	Bending (b)	Tension (m)
$(x/\rho)_0$	0.10	0.12	0.12
q_1	0.081	0.080	0.060
q_2	0.919	0.920	0.940
q_3	3.0	2.5	2.5
q_4	0.8	0.8	0.8
q_5	0.6199	1.8240	1.7872
q_6	−0.2210	−0.1340	−0.1105
q_7	−0.1541	−0.1805	−0.2309
q_8	−0.1939	−0.2011	−0.1441
q_9	−0.2346	−0.3214	−0.4331
q_{10}	1.7375	3.1312	2.7871
q_{11}	0.1502	0.0567	0.0800
q_{12}	0.0081	0.0052	0.0016

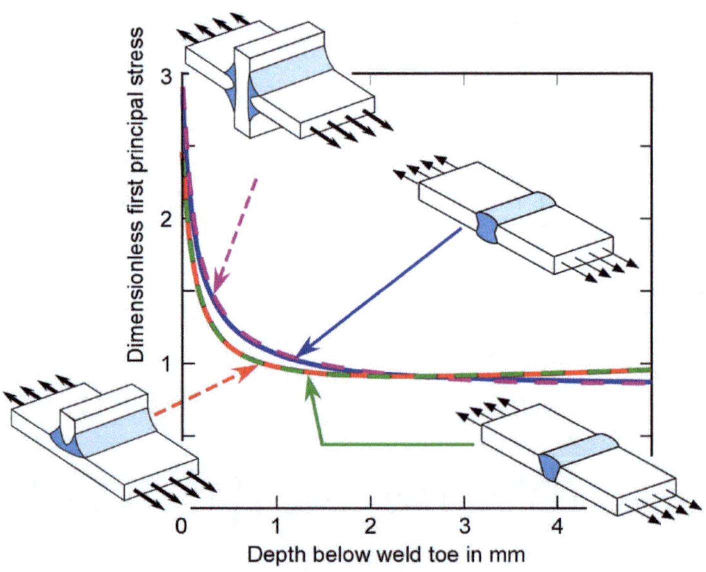

Fig. 52 Finite element based stress-depth profiles for symmetric (double V butt and cruciform) and asymmetric (V but and T joint) welds; according to Breßler (2014)

2.2.3.2.4 Approximation of the Through-Thickness Stress Profile

For some of the K-factor solutions introduced in the next step the through-thickness stress profiles of *Step 4* have to be provided in polynomial format:

$$\sigma(z) = \sum_{i=0}^{n} \sigma_i \cdot \left(\frac{z}{T}\right)^i = \sigma_0 + \sigma_1 \cdot \left(\frac{z}{T}\right) + \sigma_2 \cdot \left(\frac{z}{T}\right)^2 + \cdots + \sigma_n \cdot \left(\frac{z}{T}\right)^n \quad (35)$$

with $\sigma_i = \sigma_0, \sigma_1, \ldots, \sigma_n$ are the polynomial coefficients. Note that the quality of the polynomial approximation should always be tested, e.g., by eye.

2.2.3.2.5 Determination of Stress Intensity Factors K_A and K_C

Stress intensity factors must be separately determined for the center (A) and the surface points (C) of the semi-elliptical surface cracks in order to enable the individual determination of the depth and length growth of the crack and the crack aspect ratio a/c. Various analytical approaches can be applied to K determination:

(a) The most common approach corrects the K solution for a flat tension plate by a magnification factor M_k for the weld. This principle was first introduced by Maddox (1975) and is also part of Annex M of BS 7910 (2013), for an overview on existing solutions see Madia et al. (2017b, 2018). Although the use of these K solutions is possible within the frame of IBESS, it is not recommended as the first option. As with the stress concentration factors and through-thickness stress profile solutions (*Steps* 2 and 3), the application ranges of the M_k factor based stress intensity factor solutions are rather limited particularly with respect to small flank angles and weld toe radii.
(b) K-factors of weldments can also be obtained by the influence coefficient method in which the non-linear stress distribution is reduced to basic load cases, usually constant, linear, quadratic and cubic distributions. For each of these, stress intensity factors are determined by finite element calculations for a finite plate with semi-elliptical surface crack. Two solutions of that type (Fett et al. 1990; Shiratori et al. 1987) have been implemented within IBESS, see Madia et al. (2018).

The K-factor is then determined for a substitute geometry (Fig. 53) but in combination with the stress profile acting in the real structure which is provided in terms of Eq. (35).

$$K_I = \sqrt{\pi a} \cdot \sum_{i=0}^{n} \sigma_i F_i \cdot \left(\frac{a}{T}\right)^i = \sqrt{\pi a} \cdot \left[\sigma_0 F_0 + \sigma_1 F_1 \left(\frac{a}{T}\right) + \sigma_2 F_2 \left(\frac{a}{T}\right)^2 + \cdots + \sigma_n F_n \left(\frac{a}{T}\right)^n\right] \quad (36)$$

The influence functions F_i for the deepest and the surface points (A and C) of the semi-elliptical surface crack according to the solution of Fett et al. (1990) are provided in Tables 3 and 4. Their validity range is $0 \leq a/T \leq 0.8$ and $0 \leq a/c \leq 1$.

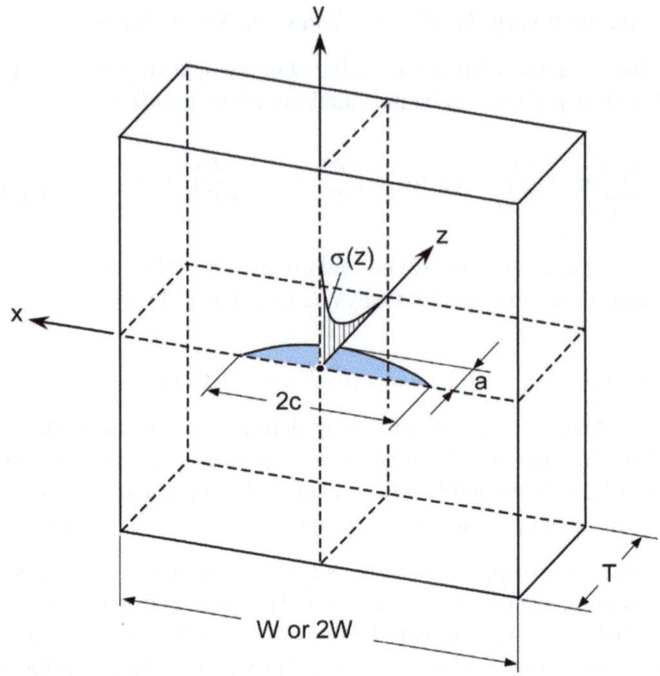

Fig. 53 Geometrical configuration used for the determination of stress intensity factors of semi-elliptical surface cracks for arbitrary stress distributions in the thickness direction

In addition, the crack dimensions should be small compared to the cross section in terms of the plate width.

Note that the component to be assessed and the substitute geometry should not be too different with respect to their stiffness. If the component is stiffer than the substitute geometry, the K-factor of the latter will be conservative. i.e., it will be over-estimated. As the consequence, tubular welds should make use of a cylindrical substitute geometry rather than of a plate as long as their wall thickness-to-diameter ratio is rather small.

An advantage of the influence functions as well as of the weight functions discussed below compared to the solutions based on the magnification factor M_k is that they are generally two-dimensional (2D) whereas some of the latter are three-dimensional (3D). This statement requires an explanation. The 2D solutions have a much wider application range since they refer to a substitute geometry which is a flat rectangular plate in most cases. In contrast, the M_k solutions have to be obtained for specific weldment geometries and they are only applicable for the range of geometrical parameters for which finite element solutions have been performed. This is restricted for reasons of effort particularly in the 3D case. One limitation has been mentioned above, namely that no solutions are available for small flank angles.

Table 3 Geometry functions F_i of Eq. (36) for the deepest point (A) of the semi-elliptical surface crack

n	a/c	a/T = 0	a/T = 0.2	a/T = 0.4	a/T = 0.6	a/T = 0.8
0	0.0	1.1225	1.3801	2.1061	4.0246	11.918
0	0.2	1.0532	1.1061	1.3051	1.5716	1.7011
0	0.4	0.9385	0.9570	1.0460	1.1460	1.1900
0	0.6	0.8330	0.8410	0.8845	0.9300	0.9600
0	0.8	0.7410	0.7460	0.7710	0.8000	0.8200
0	1.0	0.6590	0.6630	0.6780	0.6920	0.6970
1	0.0	0.6817	0.7837	1.0592	1.7504	4.4367
1	0.2	0.6063	0.6397	0.7244	0.8148	0.8796
1	0.4	0.5796	0.5945	0.6310	0.6680	0.6978
1	0.6	0.5487	0.5539	0.5680	0.5868	0.6050
1	0.8	0.5100	0.5120	0.5190	0.5310	0.5480
1	1.0	0.4705	0.4730	0.4794	0.4864	0.4965
2	0.0	0.5240	0.5820	0.7347	1.1046	2.4838
2	0.2	0.4434	0.4674	0.5245	0.5711	0.6140
2	0.4	0.4337	0.4459	0.4764	0.4954	0.5214
2	0.6	0.4254	0.4301	0.4419	0.4543	0.4755
2	0.8	0.4110	0.4125	0.4160	0.4220	0.4360
2	1.0	0.3874	0.3877	0.3896	0.3960	0.4048
3	0.0	0.4395	0.4781	0.5779	0.8137	1.6651
3	0.2	0.3570	0.3741	0.4198	0.4483	0.4807
3	0.4	0.3525	0.3633	0.3886	0.4068	0.4276
3	0.6	0.3510	0.3588	0.3707	0.3805	0.3994
3	0.8	0.3460	0.3520	0.3555	0.3620	0.3750
3	1.0	0.3365	0.3370	0.3385	0.3419	0.3487
4	0.0	0.3855	0.4135	0.4850	0.6510	1.2352
4	0.2	0.3020	0.3142	0.3480	0.3770	0.3990
4	0.4	0.3015	0.3104	0.3323	0.3500	0.3672
4	0.6	0.3005	0.3090	0.3200	0.3313	0.3461
4	0.8	0.2997	0.3060	0.3090	0.3165	0.3260
4	1.0	0.2990	0.2994	0.3000	0.3037	0.3088
5	0.0	0.3473	0.3688	0.4233	0.5477	0.9772
5	0.2	0.2685	0.2770	0.3040	0.3270	0.3430
5	0.4	0.2675	0.2733	0.2917	0.3093	0.3239
5	0.6	0.2665	0.2710	0.2850	0.2951	0.3101
5	0.8	0.2662	0.2700	0.2782	0.2840	0.2950
5	1.0	0.2660	0.2690	0.2710	0.2735	0.2777

Table 4 Geometry functions F_i of Eq. (36) for the surface points (C) of the semi-elliptical surface crack

n	a/c	a/T=0	a/T=0.2	a/T=0.4	a/T=0.6	a/T=0.8
0	0.0	0.0000	0.0000	0.0000	0.0000	0.0000
0	0.2	0.5159	0.5540	0.6549	0.8396	1.1433
0	0.4	0.6727	0.7039	0.7917	0.9212	1.1472
0	0.6	0.7232	0.7474	0.8031	0.9339	1.0702
0	0.8	0.7300	0.7490	0.7951	0.9010	0.9950
0	1.0	0.7156	0.7289	0.7767	0.8391	0.9174
1	0.0	0.0000	0.0000	0.0000	0.0000	0.0000
1	0.2	0.0685	0.0762	0.0990	0.1571	0.2427
1	0.4	0.1043	0.1138	0.1391	0.1825	0.2442
1	0.6	0.1175	0.1246	0.1449	0.1802	0.2178
1	0.8	0.1240	0.1260	0.1440	0.1670	0.1930
1	1.0	0.1178	0.1229	0.1331	0.1483	0.1668
2	0.0	0.0000	0.0000	0.0000	0.0000	0.0000
2	0.2	0.0171	0.0219	0.0390	0.0628	0.0990
2	0.4	0.0322	0.0382	0.0530	0.0739	0.0973
2	0.6	0.0392	0.0439	0.0564	0.0721	0.0870
2	0.8	0.0410	0.0455	0.0542	0.0655	0.0763
2	1.0	0.0414	0.0446	0.0503	0.0579	0.0662
3	0.0	0.0000	0.0000	0.0000	0.0000	0.0000
3	0.2	0.0086	0.0114	0.0190	0.0324	0.0552
3	0.4	0.0152	0.0183	0.0265	0.0382	0.0521
3	0.6	0.0192	0.0219	0.0290	0.0372	0.0470
3	0.8	0.0210	0.0230	0.0282	0.0332	0.0416
3	1.0	0.0216	0.0229	0.0255	0.0293	0.0350
4	0.0	0.0000	0.0000	0.0000	0.0000	0.0000
4	0.2	0.0053	0.0074	0.0121	0.0201	0.0343
4	0.4	0.0088	0.0107	0.0159	0.0241	0.0315
4	0.6	0.0114	0.0135	0.0176	0.0230	0.0286
4	0.8	0.0130	0.0138	0.0173	0.0205	0.0255
4	1.0	0.0136	0.0143	0.0154	0.0178	0.0223
5	0.0	0.0000	0.0000	0.0000	0.0000	0.0000
5	0.2	0.0038	0.0051	0.0078	0.0133	0.0226
5	0.4	0.0058	0.0071	0.0109	0.0168	0.0209
5	0.6	0.0078	0.0095	0.0120	0.0157	0.0197
5	0.8	0.0095	0.0100	0.0120	0.0145	0.0172
5	1.0	0.0096	0.0101	0.0106	0.0119	0.0153

There is, however, also a weakness of the solutions based on 2D through thickness stress profiles in that they implicitly assume the same stresses at any position in x direction (see Fig. 53). This is usually no problem for butt welds, T joints or cruciform joints. The assumption is, however, wrong with respect to joints with longitudinal stiffeners (longitudinal gussets) where additional conservatism is introduced when the stresses of the highest loaded section are taken into account.

(c) Besides the solutions of Fett et al. (1990, shown above) and that of Shiratori et al. (1987), a weight function solution according to Wang and Lambert (1998) is implemented in IBESS. The general expression of this is

$$K = \int_0^a [\sigma(z) \cdot m(z, a)]\, dz, \tag{37}$$

i.e., the stress intensity factor is determined as the product of the stress distribution σ(z) and the weight function m (z, a). Separate solutions have to be applied for the deepest (A) and the surface points (C) of the semi-elliptical surface crack.

Deepest Point of the Crack, A

The weight function solution m^A is given by

$$m^A(z, a) = \frac{2}{\sqrt{2\pi(a - z)}}\left[1 + M_1^A\left(1 - \frac{z}{a}\right)^{1/2} + M_2^A\left(1 - \frac{z}{a}\right) + M_3^A\left(1 - \frac{z}{a}\right)^{3/2}\right] \tag{38}$$

with

$$M_1^A = \frac{\pi}{\sqrt{2Q}}\left(4Y_m^A - 6Y_b^A\right) - \frac{24}{5}, \tag{39}$$

$$M_2^A = 3 \tag{40}$$

and

$$M_3^A = 2\left(\frac{\pi}{\sqrt{2Q}}Y_m^A - M_1^A - 4\right) \tag{41}$$

The indices m and b stand for membrane and bending loading. In terms of a polynomial such as Eq. (35) they would refer to σ_0 and σ_1. The auxiliary functions for Y_m^A

$$Y_m^A = B_0 + B_1\left(\frac{a}{T}\right)^2 + B_2\left(\frac{a}{T}\right)^4 + B_3\left(\frac{a}{T}\right)^6 \tag{42}$$

are

$$B_0 = 1.0929 + 0.2581\left(\frac{a}{c}\right) - 0.7703\left(\frac{a}{c}\right)^2 + 0.4394\left(\frac{a}{c}\right)^3, \tag{43}$$

$$B_1 = 0.456 - 3.045\left(\frac{a}{c}\right) + 2.007\left(\frac{a}{c}\right)^2 + 1\Big/\left[0.147 + \left(\frac{a}{c}\right)^{0.688}\right], \quad (44)$$

$$B_2 = 0.995 - 1\Big/\left[0.027 + \left(\frac{a}{c}\right)\right] + 22\left[1 - \left(\frac{a}{c}\right)\right]^{9.953}, \quad (45)$$

and

$$B_3 = -1.459 + 1\Big/\left[0.014 + \left(\frac{a}{c}\right)\right] - 24.211\left[1 - \left(\frac{a}{c}\right)\right]^{8.071}, \quad (46)$$

those for Y_b^A

$$Y_b^A = A_0 + A_1\left(\frac{a}{T}\right)^2 + A_2\left(\frac{a}{T}\right)^4 + A_3\left(\frac{a}{T}\right)^6 \quad (47)$$

are

$$A_0 = 0.4537 + 0.1231\left(\frac{a}{c}\right) - 0.7412\left(\frac{a}{c}\right)^2 + 0.4600\left(\frac{a}{c}\right)^3, \quad (48)$$

$$A_1 = -1.652 + 1.665\left(\frac{a}{c}\right) - 0.534\left(\frac{a}{c}\right)^2 + 1\Big/\left[0.198 + \left(\frac{a}{c}\right)^{0.846}\right], \quad (49)$$

$$A_2 = 3.418 - 3.126\left(\frac{a}{c}\right) - 1\Big/\left[0.041 + \left(\frac{a}{c}\right)\right] + 17.259\left[1 - \left(\frac{a}{c}\right)\right]^{9.286} \quad (50)$$

and

$$A_3 = -4.228 + 3.643\left(\frac{a}{c}\right) + 1\Big/\left[0.020 + \left(\frac{a}{c}\right)\right] - 21.924\left[1 - \left(\frac{a}{c}\right)\right]^{9.203}. \quad (51)$$

Surface Points of the Crack, C

The weight function solution m^C is given by

$$m^C(z, a) = \frac{2}{\sqrt{\pi \cdot z}}\left[1 + M_1^C\left(\frac{z}{a}\right)^{1/2} + M_2^C\left(\frac{z}{a}\right) + M_3^C\left(\frac{z}{a}\right)^{3/2}\right] \quad (52)$$

with

$$M_1^C = \frac{\pi}{\sqrt{4Q}}\left(30F_b^C - 18F_m^C\right) - 8, \quad (53)$$

$$M_2^C = \frac{\pi}{\sqrt{4Q}}\left(60F_m^C - 90F_b^C\right) + 15 \quad (54)$$

and

$$M_3^C = -\left(1 + M_1^C + M_2^C\right) \quad (55)$$

The auxiliary functions for F_m^C

$$F_m^C = \left[C_0 + C_1\left(\frac{a}{T}\right)^2 + C_2\left(\frac{a}{T}\right)^4\right]\sqrt{\left(\frac{a}{c}\right)} \tag{56}$$

are

$$C_0 = 1.2972 - 0.1548\left(\frac{a}{c}\right) - 0.0185\left(\frac{a}{c}\right)^2, \tag{57}$$

$$C_1 = 1.5083 - 1.3219\left(\frac{a}{c}\right) + 0.5128\left(\frac{a}{c}\right)^2 \tag{58}$$

and

$$C_2 = -1.101 + \frac{0.879}{0.157 + \frac{a}{c}}, \tag{59}$$

those for F_b^C

$$F_b^C = \left[D_0 + D_1\left(\frac{a}{T}\right)^2 + D_2\left(\frac{a}{T}\right)^4\right]\sqrt{\left(\frac{a}{c}\right)} \tag{60}$$

are

$$D_0 = 1.2687 - 1.0642\left(\frac{a}{c}\right) + 1.4646\left(\frac{a}{c}\right)^2 - 0.7250\left(\frac{a}{c}\right)^3, \tag{61}$$

$$D_1 = 1.1207 - 1.2289\left(\frac{a}{c}\right) + 0.5876\left(\frac{a}{c}\right)^2 \tag{62}$$

and

$$D_2 = 0.190 - 0.608\left(\frac{a}{c}\right) + \frac{0.199}{0.035 + \frac{a}{c}} \tag{63}$$

The shape factor for elliptical crack Q is

$$Q = 1.0 + 1.464\left(\frac{a}{c}\right)^{1.65} \quad \text{for} \quad 0 \le \frac{a}{c} \le 1 \tag{64}$$

Some words are due with respect to the integration algorithm. (i) Following Moftakhar and Glinka (1992) the stress profiles and the weight functions are subdivided into intervals and approximated by piecewise linear functions. This has the advantage of avoiding numerical problems associated with the interpolation of stress gradients by higher order polynomials. (ii) A specific problem is that the weight function contains a singularity in the last integration interval. To avoid this, closed form solutions K^* are applied in IBESS for the last integration interval which is generally chosen as $0.95 \le z/a \le 1$. K is determined by

$$K^A = a \cdot \sum_{i=1}^{n-1} \sigma_i^* \cdot m(a, U_i^*) + K^{A*} \tag{65}$$

for the deepest point of the crack. For its surface points (C), the closed form solution K^* has to be applied in the first interval which is generally chosen as $0 < z/a < 0.05$.

$$K^C = a \cdot \sum_{i=2}^{n} \sigma_i^* \cdot m(a, U_i^*) + K^{C*} \tag{66}$$

for the surface points of the crack. With $u = z/a$ the K^* terms are determined as

$$K^{A*} = \frac{a}{\sqrt{2\pi a}}\Big[\sigma(u=1)\cdot\left(0.894 + 0.1\,M_1^A + 0.015\,M_2^A + 0.0025\,M_3^A\right) - \frac{\sigma(u=1)-\sigma(u=0.95)}{1-0.95}\cdot\left(0.015 + 0.0025\,M_1^A + 0.00045\,M_2^A + 0.000082\,M_3^A\right)\Big] \tag{67}$$

and

$$K^{C*} = \frac{2a}{\sqrt{\pi a}}\Big[\sigma(u=0)\cdot\left(0.447214 + 0.05\,M_1^C + 0.007454\,M_2^C + 0.00125\,M_3^C\right) - \frac{\sigma(u=0.05)-\sigma(u=0)}{0.05}\cdot\left(0.007454 + 0.00125\,M_1^C + 0.000224\,M_2^C + 0.000042\,M_3^C\right)\Big] \tag{68}$$

The terms σ_i^* and U_i^* in Eqs. (65) and (66) are defined by

$$\sigma_i^* = \frac{1}{2}[\sigma(u_i) + \sigma(u_i - 1)]\cdot(u_i - u_{i-1}) \tag{69}$$

and

$$U_i^* = u_i - \frac{[2\sigma(u_{i-1}) + \sigma(u_i)]\cdot(u_i - u_{i-1})}{3[\sigma(u_i) + \sigma(u_{i-1})]} \tag{70}$$

Note that a great advantage of both the influence coefficient and weight function methods is that they allow for highly non-linear through-thickness stress profiles.

Superposition of K-factors

A principle which can simplify K-factor determination is superposition. As long as the crack opening modes (I, II, III) are identical, K-factors can be determined by simply adding the K-factors which have been obtained separately before for individual load cases. The influence coefficient method (explained under point b) makes use of this. In this sense, the coefficients C_0 and F_0 in Eqs. (35) and (36) stand for membrane loading (i.e., tension = constant across the section), C_1 and F_1 stand for bending loading (linear across the section) and the higher coefficients stand for nonlinear stresses e.g., due to a notch. When, e.g., residual stress depth profiles are given by a polynomial, K-factors can be calculated for these and added to the applied loading stresses when both, in relation to the crack, act in the same direction.

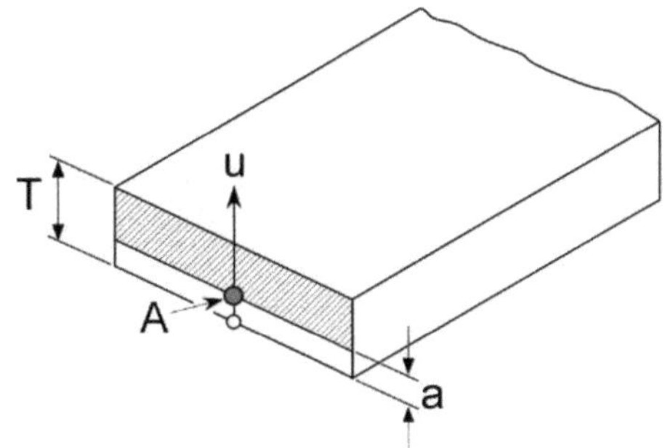

Fig. 54 Geometrical configuration used for the determination of stress intensity factors of extended surface cracks for arbitrary stress distributions in the thickness direction

Table 5 Geometry functions F_i of Eq. (71) for the extended surface crack

a/T	f_1^A	f_2^A	f_3^A	f_4^A	f_5^A
0.0	2.000	0.977	1.142	−0.350	−0.091
0.1	2.000	1.419	1.138	−0.355	−0.076
0.2	2.000	2.537	1.238	−0.347	−0.056
0.3	2.000	4.238	1.680	−0.410	−0.019
0.4	2.000	6.636	2.805	−0.611	0.039
0.5	2.000	10.02	5.500	−1.340	0.218
0.6	2.000	15.04	11.88	−3.607	0.786
0.7	2.000	23.18	28.03	−10.50	2.587
0.8	2.000	38.81	78.75	−36.60	9.871
0.9	2.000	82.70	351.0	−207.1	60.86

After a while all individual surface cracks along the weld toe have coalesced and form an extended surface crack such as shown in Fig. 54. With the c dimension of the cracks disappearing (or being infinite) the K-factor has to be determined for the crack depth only (Point A in the figure). In IBESS this is realized with a weight function solution provided by Wu and Carlsson (1991):

$$K_I = \frac{1}{\sqrt{2\pi a}} \int_0^a \sigma(u) \sum_{i=1}^{5} F_i(a/T) \cdot (1 - u/a)^{i-3/2}\, du \tag{71}$$

The geometry functions F_i^A are provided in Table 5.

2.2.3.2.6 Determination of the Reference Yield Stress σ_O

It has already been mentioned in Sect. 2.2.3.2, that the IBESS basic procedure determines the plasticity correction factor as $L_r = \sigma_{app}/\sigma_0$ for monotonic and $\Delta L_r = 0.5$

$\Delta\sigma_{app}/\sigma_0$ for cyclic loading. In principle, the use of the BS 7910 Annex P σ_{ref} solutions for $L_r = \sigma_{ref}/\sigma_Y$, the use of yield load solutions F_Y in $L_r = F/F_Y$ (Schwalbe et al. 1997, 1998), the use of predefined L_r solutions of the SRSA handbook (Dillström et al. 2008) and other formats such as provided in R6 (2014), Chapters IV.1 and VI.2 or the SINTAP compendium (Laham 1999) could be used as well. This brings up the question: Why then a new option?

(a) **Background Information**

The demands on accuracy of the model parameters K-factor and ligament plasticity factor L_r are much higher in the context of fatigue crack propagation analyses than for fracture. This is easily explained if small scale yielding conditions are assumed. In that case the crack propagation rate correlates exponentially with ΔK with the exponent being of the order of three for steels. In other words: Any error in ΔK or (ΔK_p) will be inflated on the level of the crack propagation rate and lifetime. Therefore ΔK_p (and thus L_r) should be determined as exact as possible.

The main advantage of the σ_0 based ΔL_r is its optimized accuracy. The principle is illustrated in Fig. 55 where an example is provided for the definition and determination of σ_0 for biaxial tensile loading (Madia et al. 2014), see also Zerbst et al. (2012a, 2013). Instead of being a limit stress such as in the common approaches, σ_0 is a reference yield stress adapted to the $f(L_r)$ function. Note that this is comparable with the original approach in the EPRI method (Shih and Hutchinson 1976; Kumar et al. 1981; see also Zerbst et al. 2000; Schwalbe and Zerbst 2003) and, for an overview, Sect. 2.2.9 in Zerbst and Madia (2018a) where a reference load P_0 was chosen such that, on its basis, the J-integral obtained beforehand by finite elements was reproduced.

For semi-elliptical cracks of varying size in plates loaded in tension, bending, combined tension-bending and biaxial tension the elastic plastic and the linear elastic J-integral, J and J_e, have been determined by finite elements. Based on the monotonic counterpart of Eq. (26) which can be rewritten as

$$J/J_e = [f(L_r)]^{-2} \tag{72}$$

the reference yield stress σ_0 can be determined as that stress at which $L_r = 1$. For this value of L_r the stress σ_{ref} refers to the yield strength $R_{p0.2}$ and ε_{ref} refers to the yield strain ($R_{p0.2}/E + 0.002$). When inserting these into

$$f(L_r) = \left[\frac{E \cdot \varepsilon_{ref}}{\sigma_{ref}} + \frac{1}{2}\frac{L_r^2}{E \cdot \varepsilon_{ref}/\sigma_{ref}}\right]^{-1/2}. \tag{73}$$

a material dependent value of f ($L_r = 1$) is obtained which defines the J/J_e ratio for which $\sigma_{app} = \sigma_0$. Complementary to its cyclic counterpart $\Delta\sigma_{app}$ in Eq. (27), σ_{app} is given as the applied stress referring to the gross section (without crack).

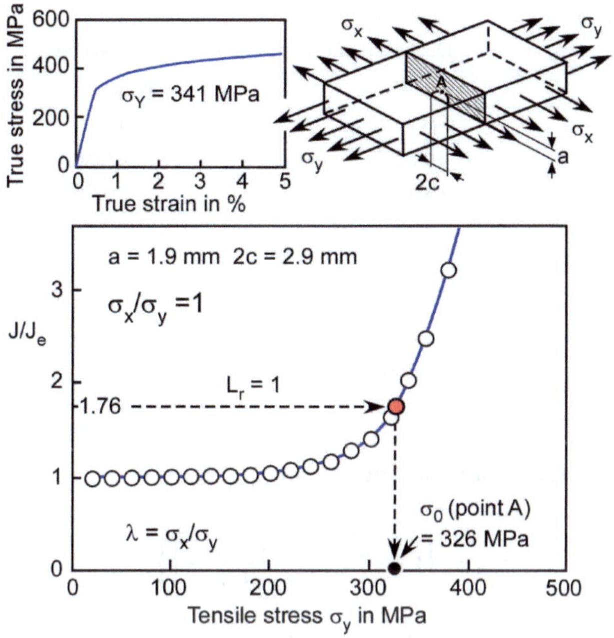

Fig. 55 Determination of the reference yield stress σ_0, Example

In the context of the IBESS method the solutions for tension, bending and combined tension-bending are relevant. With the indices m and b standing for membrane (tensile) and bending loading, the parametrized solutions for these cases are:

(b) **Parameter Equations for σ_0**

Plate Subjected to Membrane Loading ($\sigma_0 = \sigma_{0,m}$)

$$\sigma_{0,E} = \left\{ C_0 + C_1 \cdot \left[\frac{a \cdot c}{T \cdot (T + c)} \right] + C_2 \cdot \left[\frac{a \cdot c}{T \cdot (T + c)} \right]^2 \right\} \cdot \sigma_Y \tag{74}$$

with

$$C_0 = \min\left(C_{00} + \frac{\sigma_Y}{C_{01}}, 1 \right) \tag{75}$$

Validity range: $2 \cdot 10^{-4} < \frac{a \cdot c}{T \cdot (T+c)} < 0.3644$

The coefficients C_{00}, C_{01}, C_1 and C_2 are given in Table 6.

Table 6 Coefficients of Eqs. (74) and (75)

	Deepest point of the crack	Surface points
C_{00} [–]	0.9245	0.9491
C_{01} [MPa]	5419	5863
C_1 [–]	−0.5297	−0.3050
C_2 [–]	0.3863	−0.5000

Table 7 Coefficients of Eq. (76)

	Deepest point of the crack	Surface points
C_0 [–]	1.4374	1.5966
C_1 [–]	−1.2067	−0.6823

Plate Subjected to Bending Loading ($\sigma_0 = \sigma_{0,b}$)

$$\sigma_{0,b} = \left\{C_0 + C_1 \cdot \left[\frac{a \cdot c}{T \cdot (T + c)}\right]\right\} \cdot \sigma_Y \tag{76}$$

Validity range: $2 \cdot 10^{-4} < \frac{a \cdot c}{T \cdot (T+c)} < 0.2218$

The coefficients C_1 and C_2 are given in Table 7.

Plate Subjected to Combined Tension-Bending Loading

When the plate, in addition to tension (σ_m), is also subjected to bending (σ_b) the consequence will be a lowering of the reference yield strength for membrane loading $\sigma_{0,m}$ and vice versa. The reduction factor is obtained by an equation

$$\frac{\sigma_b}{\sigma_{0,b}} = 1 + C_1 \cdot \left(\frac{\sigma_m}{\sigma_{0,m}}\right) + C_2 \cdot \left(\frac{\sigma_m}{\sigma_{0,m}}\right)^2 - (1 + C_1 + C_2) \cdot \left(\frac{\sigma_m}{\sigma_{0,m}}\right)^3 \tag{77}$$

with

$$C_1 = C_{10} + C_{11} \cdot \left[\frac{a \cdot c}{T \cdot (T + c)}\right] \tag{78}$$

and

$$C_2 = C_{20} + C_{21} \cdot \left[\frac{a \cdot c}{T \cdot (T + c)}\right]\left[\frac{a \cdot c}{T \cdot (T + c)}\right]. \tag{79}$$

The coefficients C_{10}, C_{11}, C_{20} and C_{21} are given in Table 8.

Validity range: $2 \cdot 10^{-4} < \frac{a \cdot c}{T \cdot (T+c)} < 0.2218$

The principle of the correction is provided in Fig. 56.

Table 8 Coefficients of Eqs. (77)–(79)

	Deepest point of the crack	Surface points
C_{10} [–]	0.0167	0.0113
C_{11} [–]	−1.8496	−0.4266
C_{20} [–]	−1.5310	−0.9405
C_{21} [–]	3.5736	2.8513

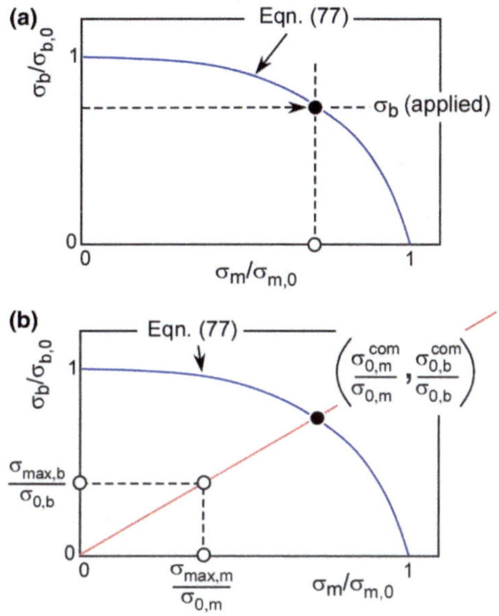

Fig. 56 Determination of the reference yield load σ_0 for combined tension and bending loading; **a** non-proportional loading; **b** proportional loading

2.2.3.2.7 Determination of the Cyclic Ligament Yielding Parameter ΔL_r

Using the reference yield stress of *Step 6*, the cyclic plasticity correction parameter is determined by Eq. (27)

2.2.3.1.8 Determination of the Plasticity Correction Function $f(\Delta L_r)$

The plasticity correction function is determined by Eq. (28).

2.2.3.2.9 Determination of ΔJ

The cyclic J-integral is obtained by Eq. (26).

2.2.3.2.10 Determination of ΔK_p

Finally, ΔJ is formally transferred to ΔK_p by Eq. (29) for its use in IBESS fatigue crack propagation analysis.

2.2.4 IBESS Procedure for Crack Closure Correction

2.2.4.1 Fatigue Crack Propagation and Crack Closure Function U

By now the crack driving force has been determined without any correction for crack closure. The effect is usually provided by a crack closure term U such that the effective stress intensity factor range is written as $\Delta K_{eff} = U \cdot \Delta K$, see Sect. 1.2.1. Based on this convention, fatigue crack propagation, within IBESS, is described by

$$\frac{da}{dN} = C \cdot \left[U \cdot \Delta K_p\right]^n \cdot \left[1 - \frac{\Delta K_{th}}{\Delta K_p}\right]^p \qquad (80)$$

with U being

$$U = \begin{cases} U_{SC}(a) & \textit{physically short}\,\text{cracks} \\ U_{LC} = \left[(1-f)/(1-R)\right] & \textit{long}\,\text{cracks} \end{cases} . \qquad (81)$$

The indices SC and LC describe the (*physically*) *short* and *long* crack stages. Whereas U_{LC} is independent of the crack size, U_{SC} (a) decreases during crack propagation such as illustrated in Fig. 57.

Note that, in Fig. 57, U_{SC} at the beginning of crack propagation is assumed to be 1, i.e., no crack closure effect has happened by then. This is certainly correct with respect to the crack closure mechanisms discussed in Sects. 1.2.2–1.2.4, i.e., the plasticity-, roughness- and oxide debris-induced ones. There is however, also the "geometric" effect mentioned at the beginning of Sect. 1.2.1. Even without the other mechanisms U should be smaller than 1 for loading at $R < 0$. This condition is fulfilled

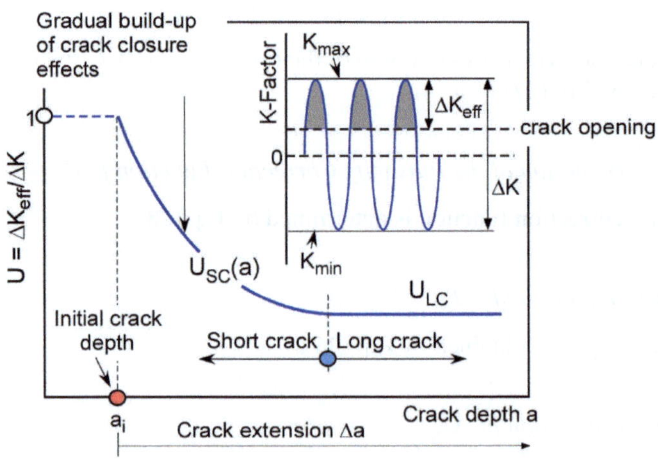

Fig. 57 Schematic crack size dependency of the dimensionless closure parameter U

for the *long* crack U_{LC} and also at the transition point from U_{SC} to U_{LC} but not in the first part of the $U_{SC}(a)$ function starting at $\Delta a=0$. Note that $U_{SC}=1$ at $a=a_i$ is chosen as a conservative option since the geometrical crack closure conditions are rather undefined for very *short* cracks.

2.2.4.2 Crack Closure Function U_{LC} for Long Cracks

With respect to *long* cracks, IBESS follows Newman's approach, also referred to as NASGRO curve (Newman 1984, NASGRO 2000). As given in Eq. (81), $U_{LC}=\left[(1-f)/(1-R)\right]$ with the function f being

$$f=\frac{\sigma_{open}}{\sigma_{max}}=\begin{cases}A_0+A_1R+A_2R^2+A_3R^3 & \text{for} \quad R\geq 0\\ A_0+A_1R & \text{for} \quad -2<R<0\end{cases} \tag{82}$$

and the coefficients A_i are

$$A_0=\left(0.825-0.34\alpha_g+0.05\alpha_g^2\right)\left[\cos\left(\frac{\pi/2\cdot\sigma_{max}}{\sigma_f}\right)\right]^{(1/\alpha_g)} \tag{83}$$

$$A_1=\left(0.415-0.071\alpha_g\right)\frac{\sigma_{max}}{\sigma_f} \tag{84}$$

$$A_2=1-A_0-A_1-A_3 \tag{85}$$

$$\text{and } A_3=2A_0+A_1-1. \tag{86}$$

Different to the original approach, the σ_{max}/σ_f ratio in Eqs. (83) and (84) is replaced by a term K_{max}/K_0 according to a slightly modified proposal of McClung (1994) which enables a wider application range of Eq. (82) which was originally obtained for tensile loading.

$$\frac{K_{max}}{K_0}=\frac{K_{max}}{\sigma'_Y\sqrt{\pi a}} \tag{87}$$

In Eqs. (83) and (84) σ_f (which in the original NASGRO curve is the so-called flow stress, i.e. the average of the monotonic uniaxial yield and tensile strengths) is replaced by the cyclic yield strength σ'_Y.

In order to avoid a negative expression in the cos (…) term of Eq. (83) a minor modification was made replacing Eq. (83) by

$$A_0=\left(0.825-0.34\alpha_g+0.05\alpha_g^2\right)\cdot\left|\cos\left(\frac{\pi\cdot\sigma_{max}}{2\sigma_f}\right)\right|^{1/\alpha_g}\cdot\text{sign}\left|\left[\cos\left(\frac{\pi\cdot\sigma_{max}}{2\sigma_f}\right)\right]\right. \tag{88}$$

2.2.4.3 Background Information About the NASGRO da/dN–ΔK Curve

Some background information is due with respect to Eqs. (82)–(86):

(a) The equations have been derived as an approximation of results obtained by an extended Dugdale or strip yield model introduced by Newman (1981) and implemented in the computer code FASTRAN (Newman 1992). The basic model is shown in Fig. 58. The crack tip region comprises three areas designated by (1), (2) and (3). Area (1) is an elastic continuum. Area (2) is a strip ahead of the crack tip which models the plastic zone. It consists of bar elements of perfect plastic material. When the crack propagates, these bars are cut but they remain at the crack faces of the crack wake in area (3). Different to the intact bars in area (2), the split bars can only transmit compressive but no tensile stresses. Let's assume a tensile overload during crack propagation. The bars in area (2) are stretched more than at the lower load cycles before. Due to crack propagation the elongated bars are "shifted" into the crack wake where they cause an increase in the crack opening stress or K-factor. Vice versa, a compression overload causes the contraction of the bars both in areas (2) and (3) and lowers the subsequent crack opening stress or K. This way the crack closure phenomenon is modeled, however, only the plasticity-induced effect, see Sect. 1.2.2. As the consequence, da/dN–ΔK curves for different R ratios and also a closure corrected da/dN–ΔK curve can be obtained by the modified strip yield model respectively its approximation by Eqs. (82)–(86) as long as the plasticity-induced mechanism dominates. This is usually the case in the Paris regime of the da/dN–ΔK curve, see Sect. 1.4.1 (Region 2 in Fig. 5) but, depending on the material, not necessarily also in the threshold region of this curve.

(b) Like the original Dugdale model, the modification was obtained for the semi-infinite tension plate with a center crack. That is the reason for McClung's modification (Eq. 87) replacing the stresses by K-factors, which extends the application range to loading types other than tension.

(c) There is, however, an even more fundamental point: the two-dimensionality of the Dugdale model requires the introduction of a constraint factor α_g to account for the three-dimensional reality. Note that this is common practice in fracture mechanics where a value of 1 is used for plane stress and a value of 3 for plane strain conditions.

Although a mechanical definition of α_g is available (Newman 1984)

$$\alpha_g = \frac{1}{A_T} \sum_{m=1}^{M} \left(\frac{\sigma_{yy}}{\sigma_0} \right)_m A_m \tag{89}$$

(with A_m being the projected area of a yielded element m on the uncracked ligament, σ_{yy}/σ_0 being the normalized crack opening stress for the element m, and A_T the total projected area of all elements (M) which have yielded)—for applications see Beretta et al. (2006), Zerbst et al. (2012), and, in more detail, Madia (2008)—its determination usually follows another way in that α_g is fitted to constant amplitude da/dN–ΔK curves determined at different load ratios R.

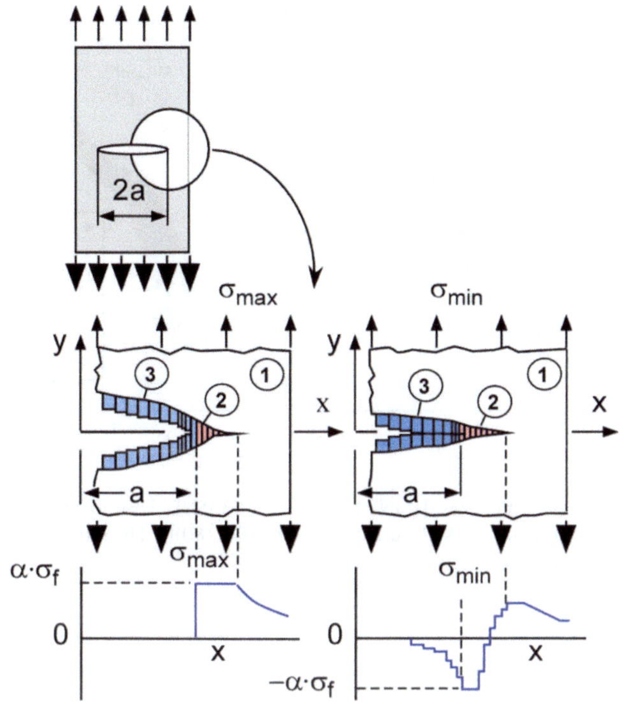

Fig. 58 Modified strip yield model. Equations (82)–(86) are based on results obtained by this

Note that uncorrected da/dN–ΔK curves show an R ratio dependency in that a higher R ratio (corresponding to a higher mean stress) increases the crack propagation rate for a given ΔK. This is explained by a lower (plasticity-induced) crack closure effect at high R. If the crack closure effect is factored out, e.g., by Eqs. (82)–(86), all curves should be fitted correctly or, if plotted as da/dN versus ΔK_{eff}, should coincide at a common upper bound provided by the closure free da/dN–ΔK curve. Since α_g is the only free parameter, it can be adapted such that it is used as a fitting parameter to optimize the regression of the fatigue crack propagation data at different R ratios. This is the basic approach adopted in the NASGRO computer code and it is also taken over by the IBESS basic procedure.

Examples for the two steels investigated are shown in Figs. 59 and 60. Note that only the base metal data have been determined since little effect of the microstructure is expected in the Paris regime. In both cases optimum results were obtained for $\alpha_g = 2.5$. Figure 59 is an example for the case that the curves corresponding to different R ratios coincide in the Paris range but not so in the threshold range. This is not surprising since, as noted above, only the plasticity-induced crack closure effect is covered by the NASGRO equation whilst other effects such as the roughness-induced one frequently dominate this region of the da/dN–ΔK curve.

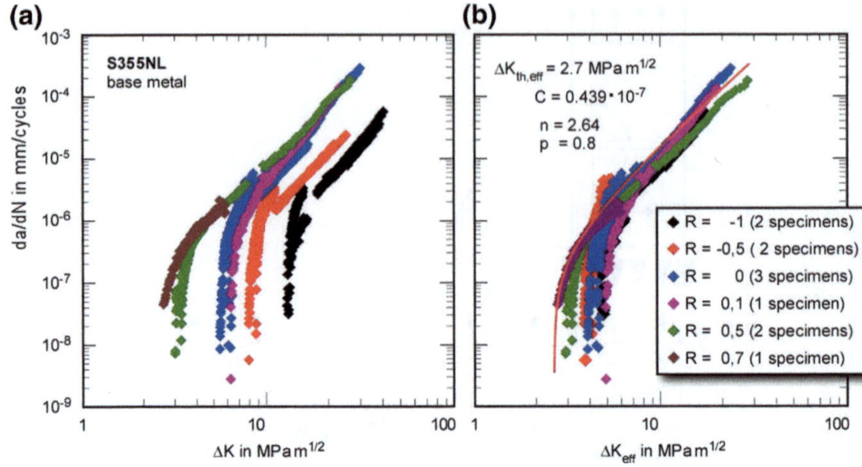

Fig. 59 Crack propagation curves da/dN–ΔK for steel S355NL, base metal. **a** Data for different R ratios; **b** data corrected for crack closure according to Eq. (80)ff. The intrinsic threshold $\Delta K_{th,eff}$ is used for ΔK_{th} in Eq. (80)

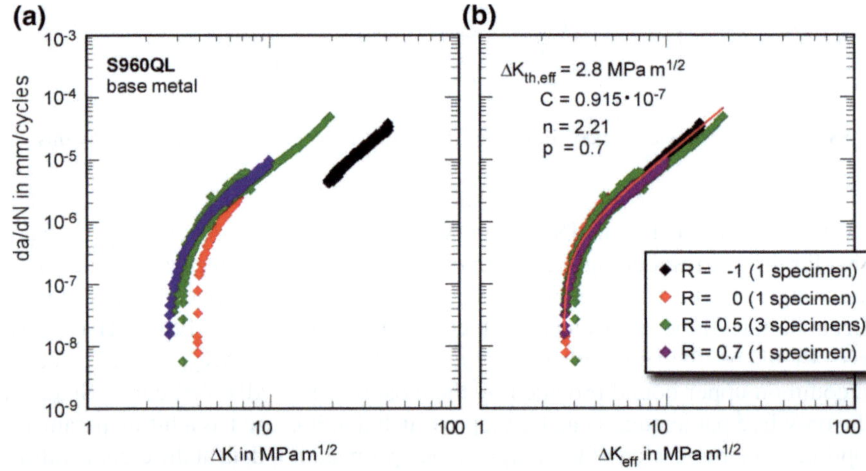

Fig. 60 Crack propagation curves da/dN–ΔK for steel S960QL, base metal. **a** Data for different R ratios; **b** data corrected for crack closure according to Eq. (80)ff. The intrinsic threshold $\Delta K_{th,eff}$ is used for ΔK_{th} in Eq. (80)

Note that the curve fit in the threshold regime is provided by the second term, $1 - (\Delta K_{th}/\Delta K_p)^p$ of Eq. (80), i.e., it is based on the fatigue crack propagation threshold ΔK_{th} and on the exponent p. The determination of ΔK_{th} has been discussed in Sect. 1.4.2. The exponent p is fitted such that the curve starts at $\Delta K_{th,eff}$ and then passes into the da/dN–ΔK_{eff} curve of the Paris regime. The fan out of the da/dN–ΔK curves in Fig. 59b is considered as an artefact due to premature crack closure (in load

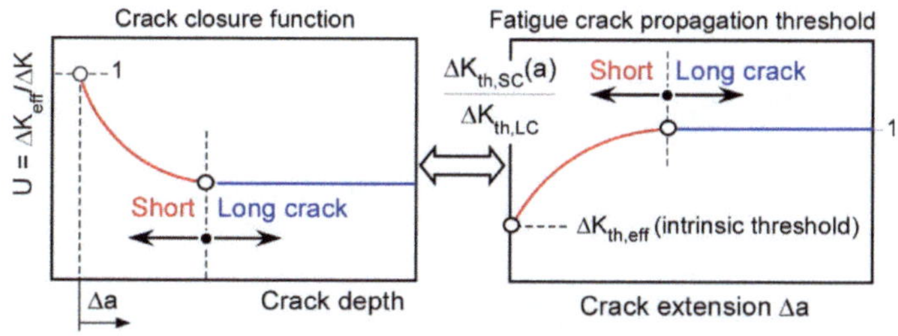

Fig. 61 Parallel development of the *short* crack closure factor U_{SC} and the *short* fatigue crack propagation threshold with increasing crack depth, schematic view

reduction) and the application of the NASGRO equations to a range where effects other than the plasticity crack closure effect play a role.

A last point to be mentioned in this context is that adequate experimental data, i.e. da/dN–ΔK curves at different R ratios will frequently be missed. In such cases, NASGRO (2000) provides fixed values for α_g, namely $\alpha_g = 2.5$ (or higher) for materials with low toughness/yield strength ratios such as high-strength steels and $\alpha_g = 1.5$–2.0 for materials with higher toughness/yield strength ratios, e.g. aluminium alloys.

2.2.4.4 Crack Closure Function U_{SC} (a) for Physically Short Cracks

It has long been recognized (Minakawa et al. 1983) that the crack closure effects and the fatigue crack propagation threshold are phenomena closely related to each other. This offers the possibility to model the development of crack closure of *physically short* cracks by the corresponding crack size dependency of the fatigue crack propagation threshold ΔK_{th}, i.e. by the cyclic R-curve, e.g. Bruzzi and McHugh (2002), McEvily et al. (2003), Chapetti et al. (2004). In IBESS, the crack closure function U_{SC} (a) for the *physically short* cracks is obtained from the long crack U_{LC} and the cyclic R-curve $\Delta K_{th,SC}$ (a) by a function

$$\frac{1 - U_{SC}(a)}{1 - U_{LC}} = \frac{\Delta K_{th,SC}(a) - \Delta K_{th,eff}}{\Delta K_{th,LC} - \Delta K_{th,eff}} \tag{90}$$

with $\Delta K_{th,LC}$ and $\Delta K_{th,eff}$ being the *long* crack and intrinsic or opening-free fatigue crack propagation thresholds. In other words: the development of the *short* crack closure function U_{SC} is mirrored by the cyclic R-curve. The *short* crack stage is terminated when the crack becomes a *long* one in both diagrams, such as illustrated in Fig. 61.

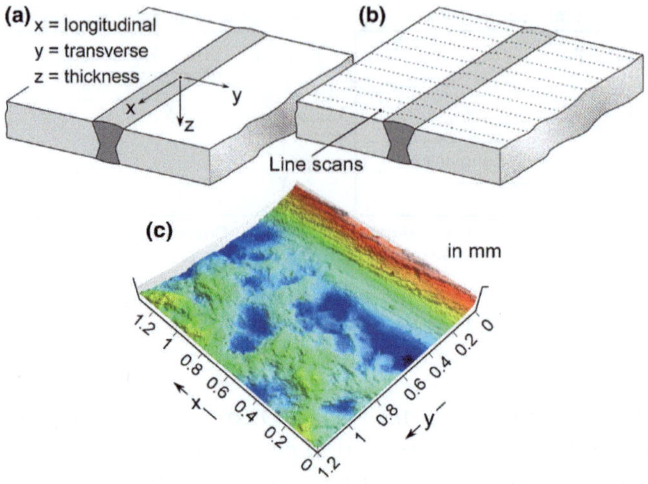

Fig. 62 Measurement of the weld seam geometry. **a** Coordinate system; **b** scanning pattern of the line scans; **c** example of a surface topography by means of optical 3D scans; according to Schork et al. (2018)

2.3 IBESS Input Data

2.3.1 Weld Toe Geometry

2.3.1.1 Brief Overview

It is known that the weld toe geometry is of paramount importance for the fatigue lifetime of a weld. The relevant geometrical parameters have already been introduced in Sect. 1.5.3, Fig. 17 together with the coordinate system applied in IBESS. It has also been shown in Sect. 1.8, Fig. 20 that there was a significant variation of two of the parameters, the weld toe radius and the flank angle along the weld toe which also showed up in the variation of the through-thickness stress profiles at different positions (Sect. 2.2.3.1, Fig. 50). Since the parameters have to be stochastically introduced in the model, see Sect. 2.4.2, they have to be provided as statistical distributions (IBESS basic procedure) or as lower (respectively upper) bounds in simplified analyses, Sects. 2.3.1.7 and 2.7.4.

Figure 62 shows the determination of the weld geometry parameters by line scans along the dotted lines within the IBESS project, for more detailed information see Schork et al. (2018) where the results of 13 configurations are provided. Note that the examples in Sects. 2.3.1.1–2.3.1.5 all refer to S355NL butt welds with a plate thickness of $T = 10$ mm

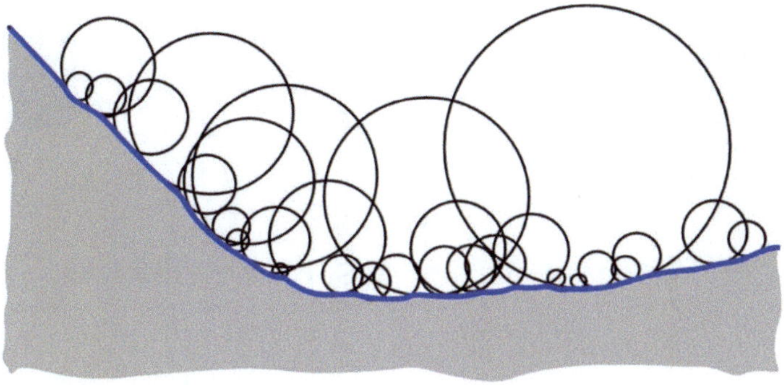

Fig. 63 Problems in defining the weld toe radius. The zoomed view shows various radii dependent on the scale to be considered; according to Lindgreen and Stenberg (2011)

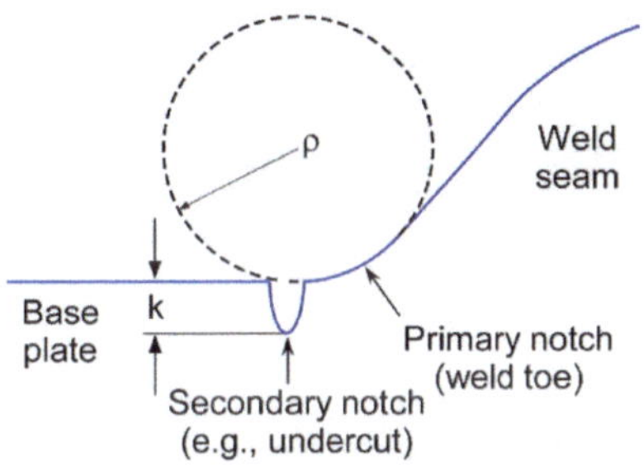

Fig. 64 Definition and separation between the weld toe radius (as the primary notch) and a secondary notch, e.g. an undercut, schematic view

2.3.1.2 Weld Toe Radius ρ

With respect to the weld toe radius, a problem of principle must be solved. Depending on the level of zooming, quite different radii can be identified at one weld toe site such as illustrated in Fig. 63. This brings up the question how to define the radius in the correct way in the context of fatigue life consideration. Note that, in spite of a large number of weld toe radii published in diverse case studies, this question is usually not addressed. Nevertheless, it is an essential one since the local notch geometry at the weld toe strongly affects the lifetime-determining *short* crack growth stage.

In IBESS, the problem is solved in that it distinguishes between the weld toe as the primary notch and, next to or inside this, secondary notches which can be undercuts, surface roughness or weld ripple edges, Sect. 2.3.1.5, Fig. 64. Thus, the weld toe radius is defined as a "global" parameter. Note that the conditions in Fig. 64 are idealized. In practical application, the weld profile will be less smooth because of which assessment by eye will usually be necessary. This is, however, not considered as critical because the scatter generated this way is rather small.

Figure 65 provides an example of a histogram and a statistical distribution of weld toe radii. As for the following geometric parameters discussed in Sect. 2.3.1 as well, a lognormal distribution is used because of the satisfying fitting of the experimental data but also in order to avoid negative percentile values. Note that the variation of

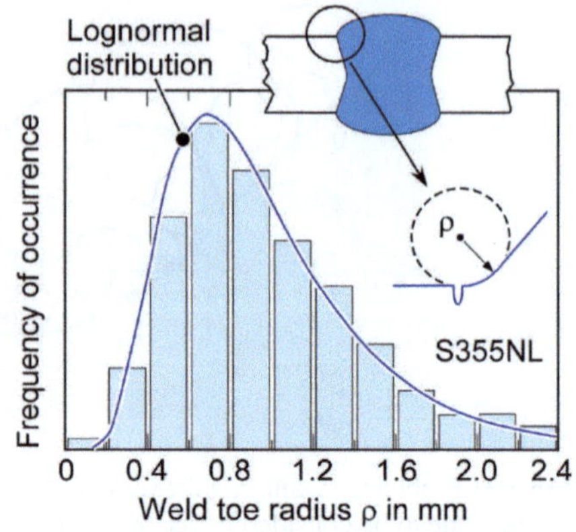

Fig. 65 Example of the statistical distribution of the weld toe radius ρ along the weld toe as obtained in IBESS; steel S355NL, butt weld; according to Schork et al. (2018)

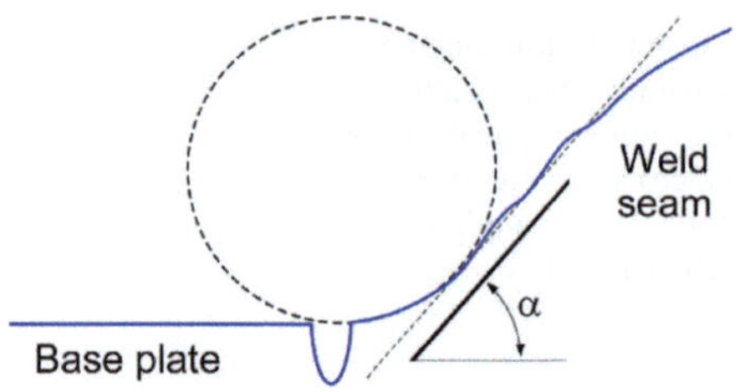

Fig. 66 Schematic view of the determination of the flank angle at the weld toe

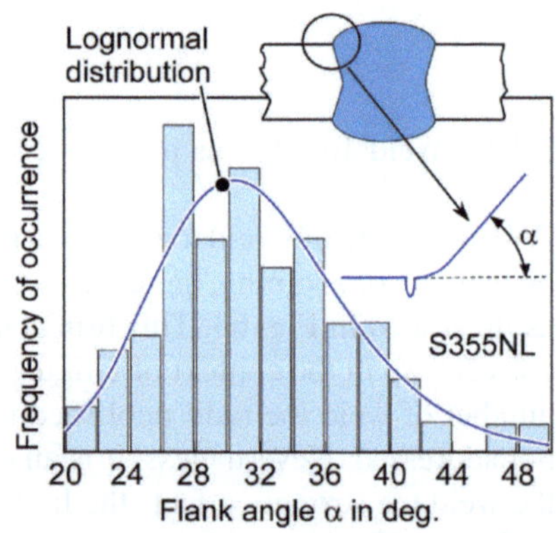

Fig. 67 Example of the statistical distribution of the flank angle α along the weld toe as obtained in IBESS; steel S355NL, butt weld; according to Schork et al. (2018)

the geometrical parameters in the IBESS analyses, is exploited only within the 10 and 90% percentile limits (Sect. 2.4.2)

2.3.1.3 Weld Toe Flank Angle α

The definition of this parameter is provided in Fig. 66. As for the toe radius, some judgement by eye will usually be necessary. A histogram and the statistical distribution based on this is shown in Fig. 67.

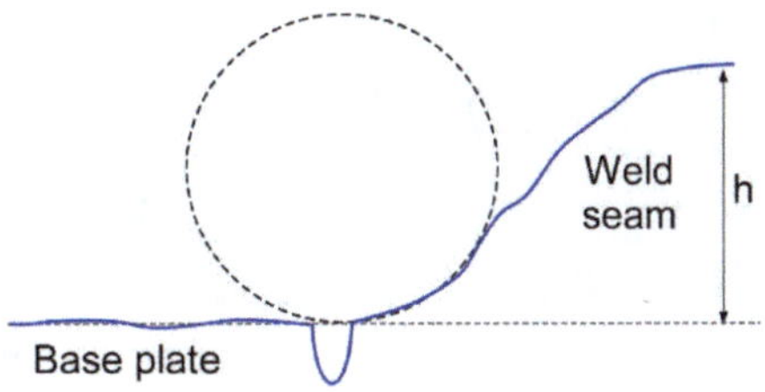

Fig. 68 Schematic view of the determination of the weld reinforcement of a butt weld

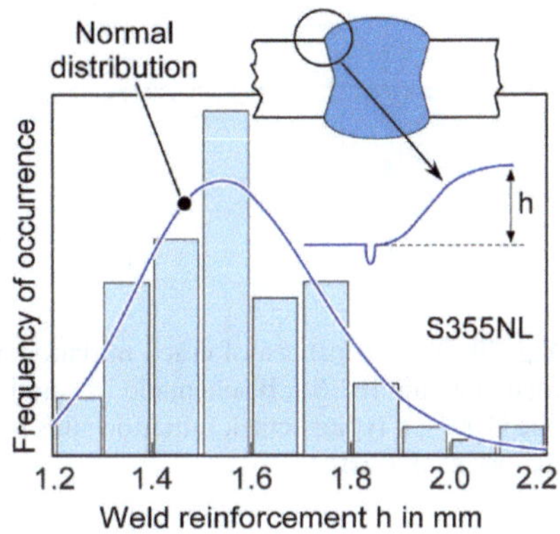

Fig. 69 Example of the statistical distribution of the weld reinforcement h of a butt weld as obtained in IBESS; steel S355NL; according to Schork et al. (2018)

2.3.1.4 Weld Reinforcement h

The definition of reinforcement is provided in Fig. 68, an example for statistical processing in Fig. 69. Note that in this case a normal distribution has been used. As mentioned above, this is permissible in IBESS as long as the 10 and 90% percentile values do not become negative.

2.3.1.5 Secondary Notch Depth k

In IBESS different crack initiation sites along the weld toe have been identified, an example is provided in Fig. 70 where cracks have been visualized by heat tinting after about one third of the overall lifetime of the weldments. It was found that cracks were initiated at base plate surface roughness features next to the weld toe (from the hot rolling process) as well as at welding ripple edges. Cracks were also initiated in the weld seam at some distance from and parallel to the toe, but these tended to arrest or soon to coalesce with the main crack, for details see Schork et al. (2018). Roughness "dimples" at the base plate of S355NL next to the weld toe are shown in Fig. 71. The topic of roughness specification close to the weld toe will be separately addressed in Annex D. Figure 72 provides the statistics of the secondary notch depth k based on roughness measurements.

With respect to the treatment of secondary notches in fracture mechanics analysis, two options are conceivable, see Zerbst et al. (2018b, c), Fig. 73.

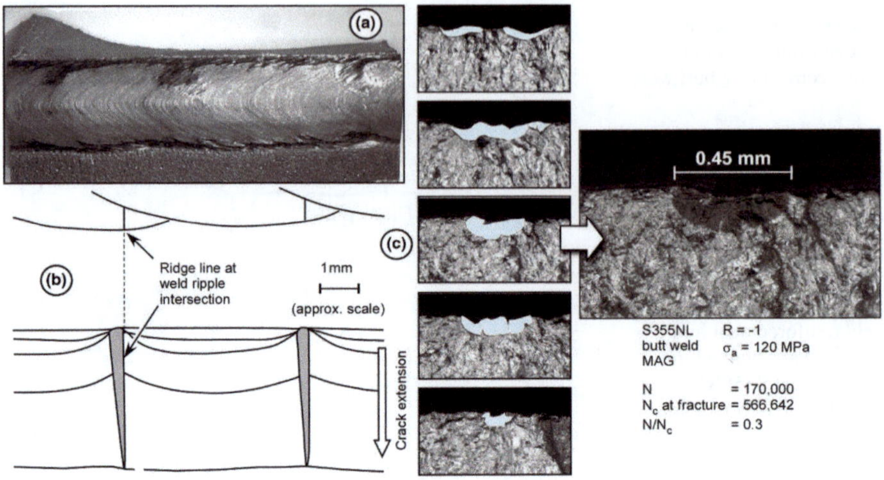

Fig. 70 Typical pattern of crack initiation and early crack growth. **a** Weld ripple structure investigated within IBESS; **b** schematic illustration of early crack evolution according to Verreman and Nie (1991); **c** typical crack initiation sites found in IBESS (marked as hatched areas); according to Schork et al. (2018)

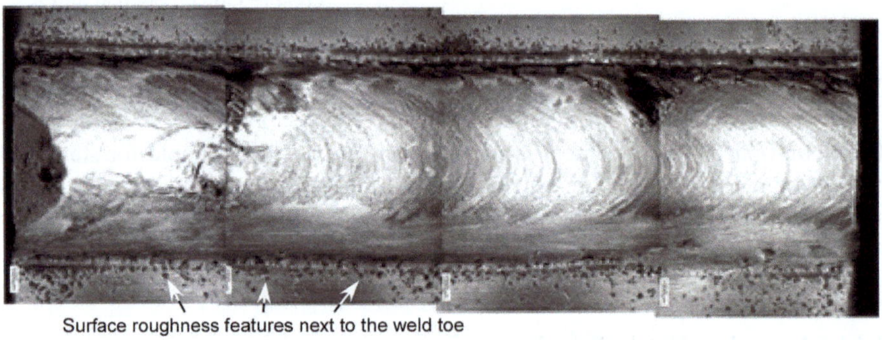

Fig. 71 Base plate surface roughness close to the fusion line of welds investigated in IBESS; according to Madia et al. (2018)

(a) The defects are treated as notches at the crack driving force side of the component. This option requires the determination of the local stress-depth profiles ahead of the secondary notches usually by finite element analyses. In IBESS, finite element based solutions of the through-thickness stress profiles available for primary notches are provided in table format for single- and double-V butt welds of 10 mm thick plates loaded in tension or bending (Zerbst 2016).

(b) The defects are considered as parts of the cracks, i.e., k contributes to the initial crack depth a_i. The realization of this option is much easier than those of option (a) because of which it is chosen for the IBESS basic procedure.

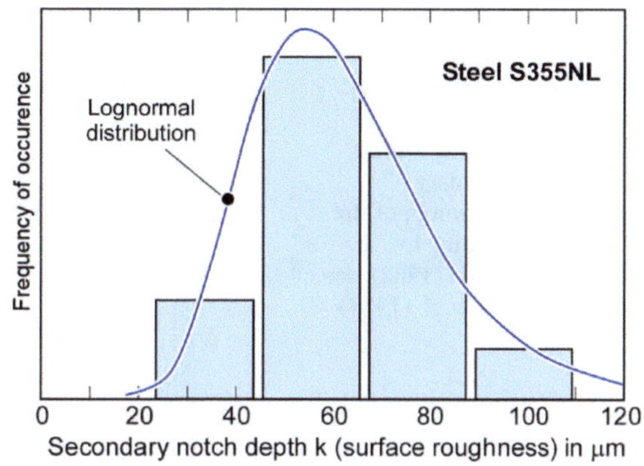

Fig. 72 Example of the statistical distribution of the secondary notch depth k (in the present case given by the base plate roughness close to the weld toe) as obtained in IBESS; steel S355NL, butt weld; according to Schork et al. (2018)

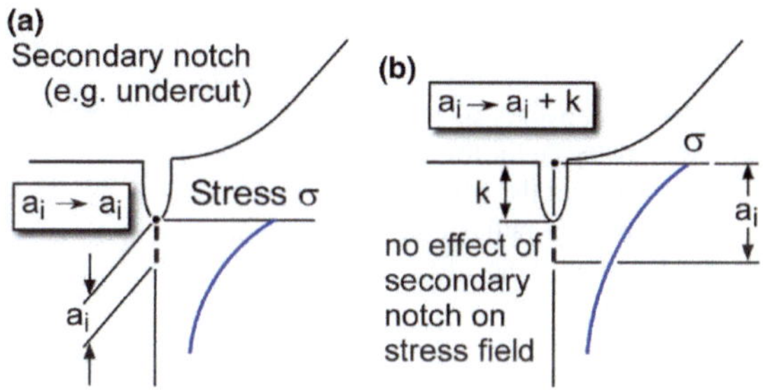

Fig. 73 Two options of the treatment of the secondary notch depth in IBESS; **a** as notch, i.e., as causing stress concentration **b** as part of the initial crack

In principle, the question which of the two options is appropriate, depends on the ratio of the crack depth ahead of the notch and the notch radius, a/ρ. Murakami (2002), based on linear-elastic considerations, concluded that a notch with a small crack at its tip may usually be regarded as a crack. Figure 74 shows the boundary correction functions F_I (referring to K_I) of cracks emanating from an elliptical hole. Values other than $F_I = 1$ are an indication of a notch effect on the stress intensity. However, when a value of 1 is approached, no difference is stated between the analyses with and without explicit consideration of the notch. In other words: adding the notch depth to the crack (option b) provides correct results of K_I in such a case. As can be seen, besides a/ρ the notch geometry, i.e., its b/k ratio also plays a role. If we accept an underestimation of K_I by no more than 2.5%, option (b) generally yields satisfying results when the crack length a exceeds the notch radius by a factor of 2.2, for narrower notches this ratio becomes even smaller (0.5 for $b/k = 4$ and 0.9 for $b/k = 2$). If a 5% error in K_I were accepted, option (b) would be generally acceptable even for a crack depth of only 0.8 times the notch root radius.

Note that these considerations are valid for an elliptical notch in an infinite plate. However, in the IBESS case, we have a multiple stress concentration problem because of the two stress raisers of the weld toe and the secondary notch. Figure 75 shows the dependency of the K factor of short cracks controlled by the notch stress field and

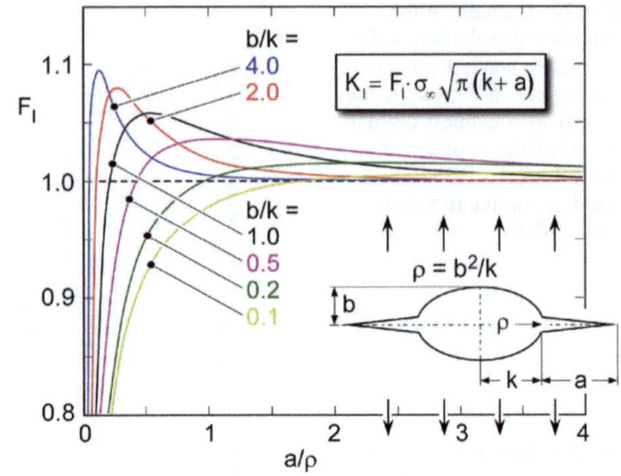

Fig. 74 Effect of the secondary notch depth and geometry as well as of the depth a of the crack emanating from its root on the mode I boundary correction function F_I of the K-factor; according to Nisitani and Isida (1982), see also Murakami et al. (1987)

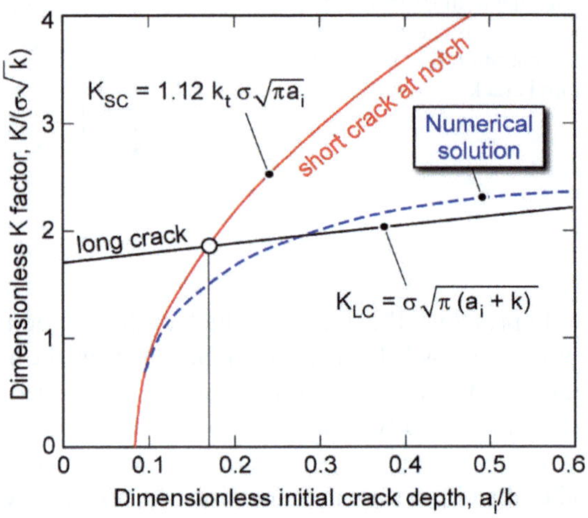

Fig. 75 Dimensionless stress intensity factor for an initial crack a_i at the edge of a circular hole of radius k in an infinite plate subject to uniaxial tensile stress σ; according to Dowling (1979), see also Suresh (2003) and Madia et al. (2018)

those of the long crack due to the remote tensile stress along with the numerical K factor as a function of the crack depth normalized with the secondary notch depth k, for details see Dowling (1979), also Suresh (2003) and Madia et al. (2018). As can be seen there is a transition crack depth $a_{i,t}$ above which the secondary notch can be treated as part of the crack. Based in the equations given in Fig. 75 this is obtained as

$$a_{i,t} = \frac{k}{\left(1.12 \cdot \mathrm{SCF}/\,\mathrm{Y}\right)^2 - 1} \tag{91}$$

Whether the notch is a blunt or a sharp one is taken into account by the elastic stress concentration factor SCF. For the deepest secondary notch depth found in IBESS in the order of k = 100 μm (Schork et al. 2018), the equation $a_i \rightarrow a_i + k$ is almost universally applicable.

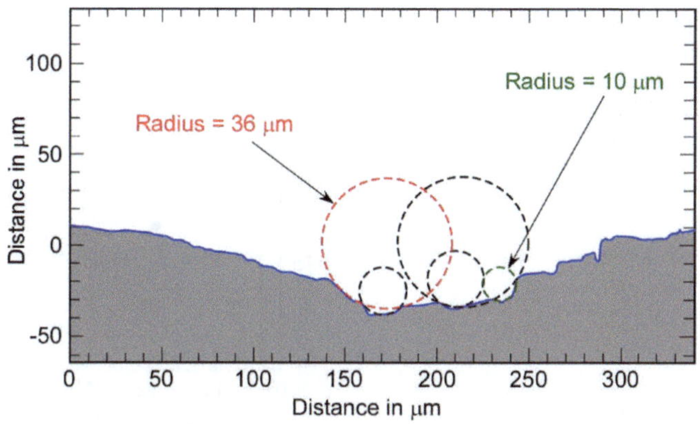

Fig. 76 Empirical notch root radius of secondary notches at the weld toe, example, steel S355NL, butt weld, data from IBESS

Note that, with respect to the notch root radius, the same problem emerges as with the toe radius in that quite different values can be obtained at slightly different positions, Fig. 76, and it is not clear beforehand at which position a crack will initiate. As the consequence, some uncertainty remains about the applicability of option (b), however, the potential error, if it occurs at all, is rather small. For further discussion see Madia et al. (2018).

In the present approach, the secondary notch depth k referring to the surface roughness of the base plates, such as statistically processed in Fig. 72, is equated with the roughness measure P_t. This is regarded as a conservative measure. For a brief discussion of this parameter see Annex D.

2.3.1.6 Further Geometry Parameters

Further geometry parameters are the plate thickness T (which shows only low variability within the manufacturing tolerances) and the weld width L. The statistical distribution of the latter is shown in Fig. 77 for a typical S-N curve data set in IBESS. Note, however, that L is not implemented as a stochastic parameter in the IBESS basic procedure for now.

2.3.1.7 Weld Geometry and Extended Weld Quality Categories

With respect to the quality standard system of weldments a change in thinking is currently to be stated. The common system follows what may be designated as a "good workmanship" philosophy. Any flaw should be avoided. Thus, a comprehensive list of different types of weld imperfections is provided in ISO 5817 (2014). These include cracks, porosity, inclusions, lack of fusion, undercuts, misalignment,

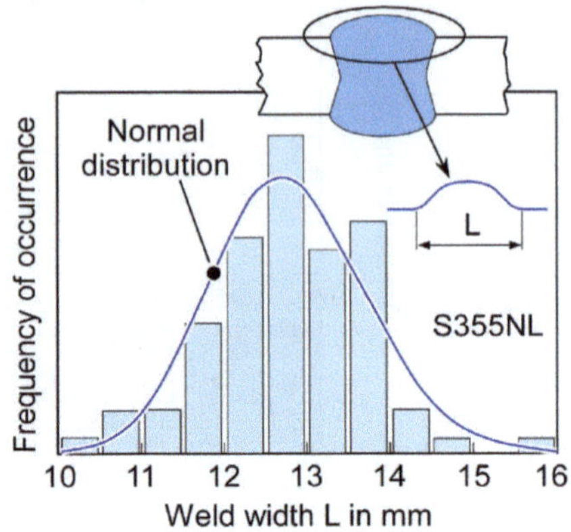

Fig. 77 Example of the statistical distribution of the weld width L as obtained in IBESS; steel S355NL, butt weld; according to Schork et al. (2018)

arc strikes and others items which, from the perspective of mechanical integrity, can be assigned to distinct groups, see Zerbst et al. (2014a):

(a) *Long* cracks and crack-like imperfections (i.e. hot and cold cracks, lack of fusion, incomplete penetration etc.): These are not permitted. If they, nevertheless, occur they are immediately subject to (*long* crack) fracture mechanics analysis.
(b) Material imperfections which act as crack initiation sites (i.e. microcracks, slag or oxide inclusions): These are of paramount importance for fatigue strength and fatigue life analyses. Depending on the quality level they are partly permissible but their sizes are restricted to certain limits.
(c) Geometric discontinuities such as misalignment (e.g., edge offset or angular distortion) which, in axially loaded joints, increase the local stresses by adding (local) bending stresses: Note that a relatively large degree of misalignment is permitted in the current guidelines which may cause considerable scatter in the corresponding S-N data. Therefore, it is suggested not to treat misalignment as a weld quality criterion but to treat it as part of the structural stress, e.g. Björk et al. (2008).
(d) Finally there are imperfections which will have no or a minor effect on the fatigue life. This is, e.g., the case with porosity or inclusions in low stress regions of the component. However, caution is advised with respect to clustered porosity or inclusions or pore/inclusion chains, particularly if they are located close to the surface.

The difficulty with the existing system is that it refers to almost any flaw regardless of whether it would negatively affect structural integrity or not. Nevertheless, the final quality level is based on the worst flaw of the weld with the consequence of unjustified conservatism in some cases. To improve this situation, proposals have been made to develop a revised weld quality system restricted to parameters which are known to influence structural integrity, particularly the fatigue strength.

The Volvo Standard STD 181-0004 (2008), see also Jonsson et al. (2011), defines three quality levels for cyclic loading (VD for normal quality, VC for high quality

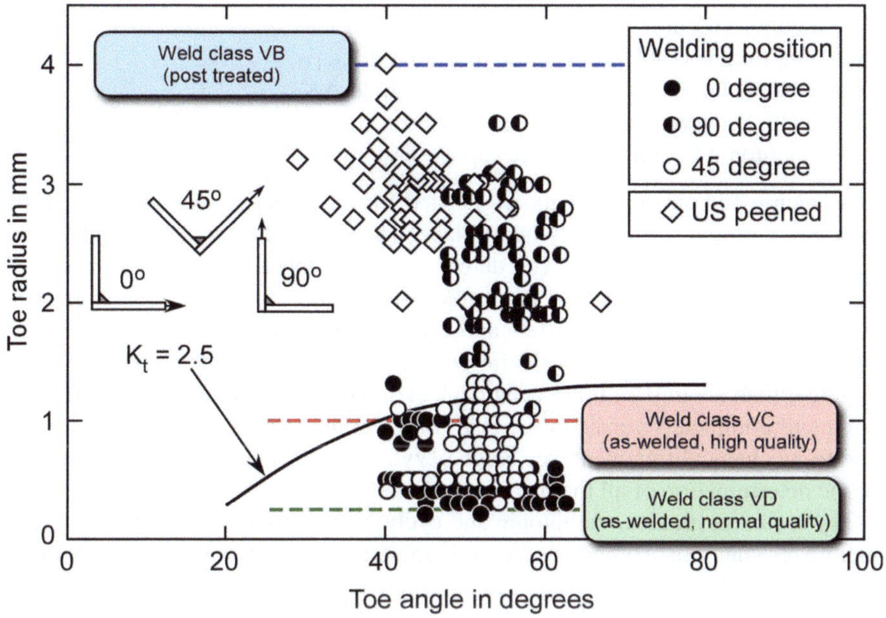

Fig. 78 VOLVO STD 181-0004 quality system: quality classes along with empirical data for weld toe radii and flank angles; according to Barsoum and Jonsson (2011)

and VB for post treated welds) and one (VS) for static loading. The underlying characteristics comprise eight parameters including the weld toe radius, the flank angle, weld reinforcement and undercuts. An example with respect to the radius is shown in Fig. 78. Likewise, the latest update of ISO 5817 (2014) contains an (informative) annex with additional requirements for fatigue. Essentially, these combine the ISO quality levels with the IIW FAT classes (for the latter see Hobbacher 2016). So far, the annex is limited to FAT 63, 90 and 125 in combination with quality levels. One of the resulting classes, B125, is partially reproduced in Table 9.

In principle, the limit values of the parameters constituting the revised weld quality categories can be used as input data to an IBESS analysis instead of the experimentally determined values of Sects. 2.3.1.2–2.3.1.6. Some limited validation will be provided in Sect. 2.7.4. Note, however, that further discussion regarding this issue is necessary.

2.3.2 Material Data

2.3.2.1 Brief Overview

The material data needed for an IBESS analysis comprise:

- the material fatigue limit,
- the cyclic stress-strain curve,

- the da/dN–ΔK curve for *long* crack propagation,
- the *long* fatigue crack propagation threshold $\Delta K_{th,LC}$,
- the intrinsic fatigue crack propagation threshold $\Delta K_{th,eff}$,
- the cyclic R-curve, and
- the monotonic fracture resistance (optional).

As mentioned above, the data are needed for the material section in which the *short* crack propagation mainly takes place. This is usually the heat affected zone (HAZ) or a specific part of it. The determination of these data is not an easy task because, depending on the heat input, the width of this zone is of the order of no more than a few millimeters or less and even over this distance a material gradient exists. For the welds under consideration within IBESS the coarse grain HAZ was not larger than about 0.5 mm (Kucharczyk et al. 2018). To circumvent the problem, specimens completely consisting of HAZ microstructure were produced by suitable thermo-mechanical treatment, see Kucharczyk et al. (2018). These were then used for the determination of all the parameters listed above.

The hardness profiles, monotonic and cyclic stress-strain curves and monotonic fracture resistance data of the two steels S355NL and S960QL and their HAZs investigated within IBESS have already been presented in Sect. 1.9.2. Figure 79 compares the HAZ microstructures in the weld with those obtained in the thermo-mechanically simulated specimens. Besides the comparison of the hardness in the weld and the substitute specimens the evaluation of the success of the thermo-mechanical simulation was based on this information.

Next to the statistical specification of the geometrical parameters of the weld toe, Sect. 2.3.1, the determination of the material data is the most expensive and, therefore, also the most limiting part of an IBESS analysis from a practical point of view. This is the reason why proposals have been developed for the simplified determination of almost all the parameters listed above. Note, however, that only parts of these have been comprehensively validated to date.

Table 9 Excerpt of ISO 5817:2014-06, Annex C (2014)

Parameter	Designation		Main document	Annex C
	Current No.	ISO 6520-1	B	B125
Excess weld metal h	1.9	502	$\leq$1 mm+0.1 L max. 5 mm	$\leq$0.2 mm+0.1 L max. 2 mm
Toe radius ρ	–	5052	–	$\geq$4 mm
Weld opening angle $180° - \alpha$	1.12	505	$\geq$150°	$\geq$150°
Axial misalignment	3.1	5071	$\leq$0.2 mm+0.1 T[a]	$\leq$ 0.05 T max. 1.5 mm
Spatter	1.23	602	Permissibility depends on the application	Not permissible

[a]For T = 0.5–3 mm, above T = 3 mm: $\leq$ 0,1 T; but max. 3 mm

Additional requirements for correlating the quality criteria with IIW-FAT classes (for butt welds with a thickness of T $\geq$ 0.5 mm)

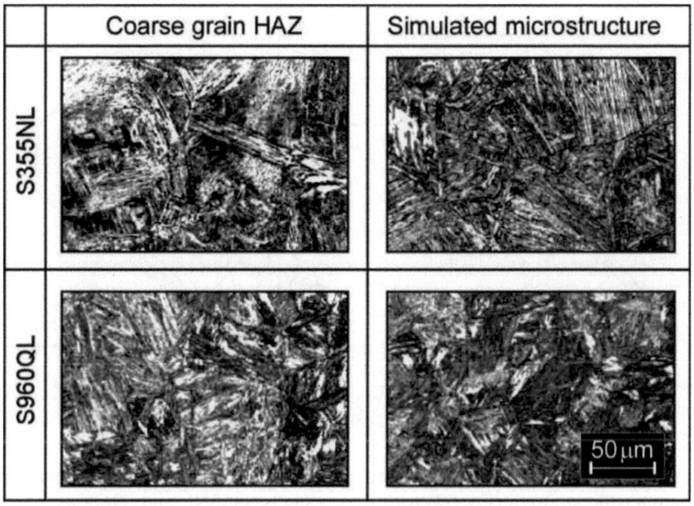

Fig. 79 Comparison of the HAZ microstructures of steels S355NL and S960QL in the coarse grain region of the welds with the thermically simulated microstructures in the substitute specimens; IBESS investigations; according to Kucharczyk (2018)

2.3.2.2 Fatigue Limit

(a) **IBESS Basic Procedure**

Not much has to be explained in general about the determination of the material fatigue limit σ_e (or its equivalent referring to, e.g., 10^7 loading cycles). The parameter is determined with smooth specimens (usually of "hourglass" type) for a load ratio of $R = -1$ (ISO 1099, 2006). As mentioned above, the material fatigue limit was determined also on specimens with thermo-mechanically simulated HAZ microstructure. Using the staircase (or "up and down") method, fatigue limits were determined as statistical parameters for the base metal and HAZ conditions of the two steels investigated, for details see Kucharczyk et al. (2018).

(b) **Simplified Approach**

A common method for the simplified determination of σ_e is the correlation with the (monotonic) ultimate tensile strength R_m or—for HAZs more suitable—the hardness. A simple correlation with R_m is provided by the German FKM Guideline for the "Analytical strength assessment of components made of steel, cast iron and aluminium materials in mechanical engineering" (Rennert et al. 2013):

$$\sigma_e(R = -1) = f_\sigma \cdot R_m \tag{92}$$

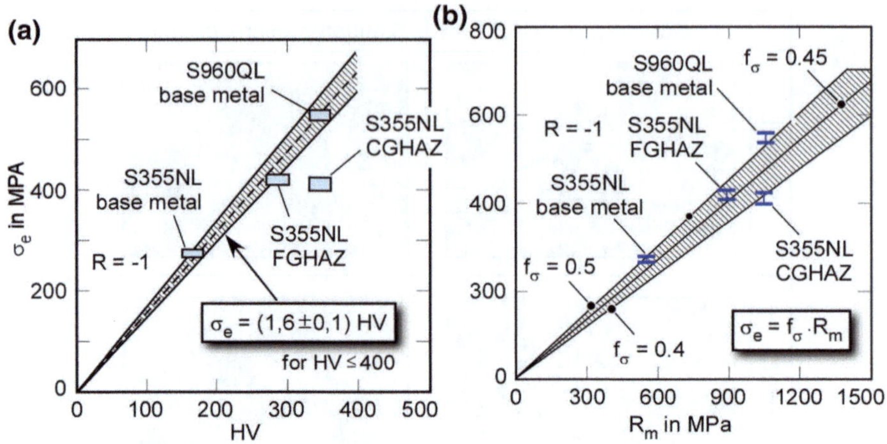

Fig. 80 Estimation of the materials fatigue limit σ_e (for $R = -1$); **a** from Vickers hardness (Eq. 93); **b** from ultimate tensile strength (Eq. 92) for the two steels S355NL and S960QL, base metals and coarse grain heat affected zones (HAZ), according to Kucharczyk et al. (2018)

with the coefficient f_σ being

- 0.4 for case hardened, austenitic stainless and forging steel (for the two latter: preliminary),
- 0.45 for other steels,
- 0.34 for steel casting, spheroidal graphite cast iron and cast iron with lamellar graphite, and
- 0.3 for malleable cast iron, wrought aluminium alloys and aluminium casting.

Estimates for further materials are provided by Radaj and Vormwald (2007). A correlation with Vickers hardness HV for steel is published in Murakami (2002)

$$\sigma_e(R = -1) = 1.6\ \mathrm{HV} \pm 0.1\ \mathrm{HV}\ \text{for}\ \mathrm{HV} \leq 400 \tag{93}$$

In Fig. 80, Eqs. (92) and (93) are applied to the two steels investigated within IBESS, i.e., its base plate and HAZ fatigue limits including the scatter. The coefficient f_σ was taken as $f_\sigma = 0.45$. As can be seen, the equations provide quite satisfying results except for the coarse grain HAZ of S355NL where the hardness correlation gave a more pronounced underestimate of σ_e.

Equation (92) can also be applied in combination with experimentally obtained R_m values from micro-tensile tests or from notched cross weld tensile tests. These were not applied within IBESS, some discussion is, however provided by Zerbst et al. (2014a).

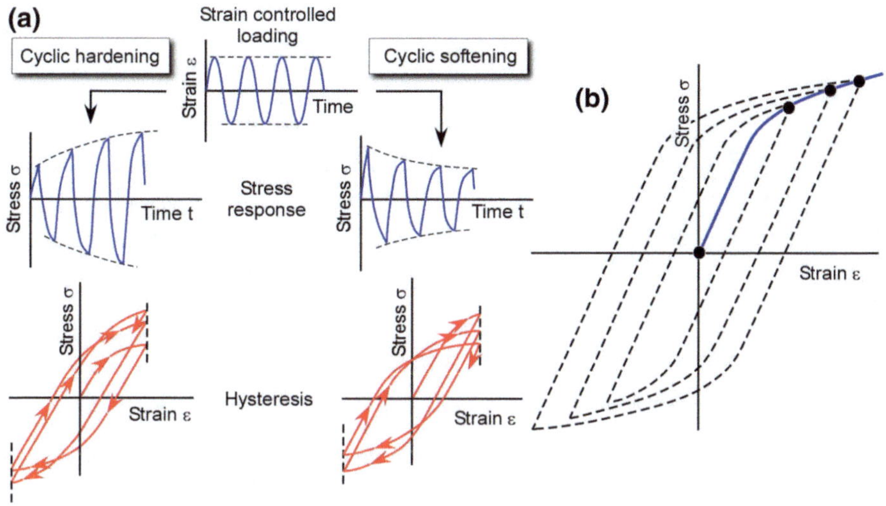

Fig. 81 **a** Cyclic hardening and softening, figure based on Landgraf (1970); **b** cyclic stress-strain curve

2.3.2.3 Cyclic Stress-Strain Curve

When subjected to repeated plastic strains, metallic materials may experience what is designated by "cyclic hardening" and "cyclic softening". Cyclic hardening means that the yield strength and the corresponding region of the stress-strain curve increases under strain controlled cyclic loading whereas cyclic softening describes the opposite, i.e., the yield stress decreases (Fig. 81). Usually after some ten or hundred loading cycles a stabilized state is reached.

(a) **IBESS Basic Procedure**

Cyclic stress-strain curves are determined with unnotched specimens at a load ratio of R = −1 (ISO 12106, 2017). The stress amplitudes used for determining the cyclic stress-strain curves are usually measured at half lifetime with respect to the number of loading cycles at fracture. In IBESS, the curves were determined in so-called single step tests where multiple test pieces are tested at different fixed strain amplitudes. The alternative consists of incremental step tests where the strain amplitude is repeatedly incrementally increased every half block cycles and, when a certain maximum strain is reached, decreased again. The results have already been shown in Sect. 1.9.2.2. It showed up that all investigated steels, i.e., their base metals and different HAZ regions, except S355NL base material experienced cyclic softening. An example is provided in Fig. 82.

For the use within the IBESS approach the cyclic stress-strain curve is fitted by a Ramberg-Osgood expression

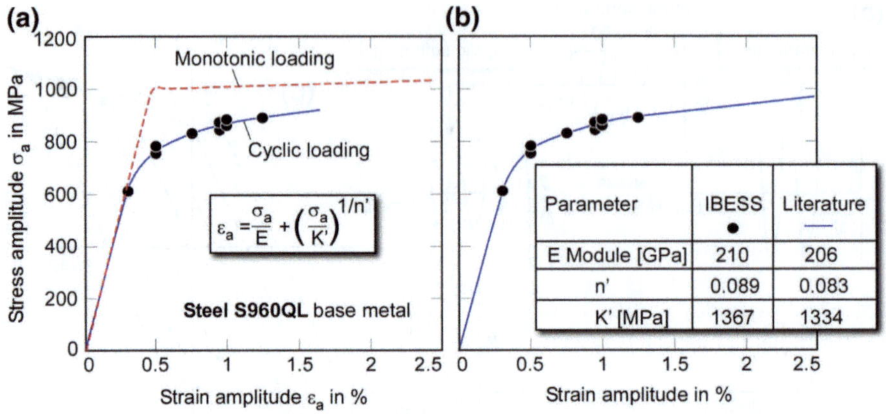

Fig. 82 Comparison **a** between monotonic and cyclic stress-strain curves of steel S960QL, base metal and **b** with literature data of a similar steel (Pusch and Hübner 1998)

$$\varepsilon_a = \frac{\sigma_a}{E} + \left[\frac{\sigma_a}{K'}\right]^{1/n'} \tag{94}$$

for the stress and strain amplitudes or by

$$\Delta\varepsilon = \frac{\Delta\sigma}{E} + 2\left[\frac{\Delta\sigma}{2K'}\right]^{1/n'} \tag{95}$$

for the stress and strain ranges when either branch of the cyclically stabilized stress-strain loop is geometrically similar to the Ramberg-Osgood equation by a scale factor of 2 (Masing's hypothesis).

(b) **Simplified Approach**

Various equations have been proposed for estimates of the parameters K′ and n′ in the literature (for those based on monotonic stress-strain curve data, see Marohnic et al. 2017). With

$$\sigma'_Y = K'(0.002)^{n'} \tag{96}$$

which is obtained when the plastic part of Eq. (94)—i.e., the term σ_a/E is omitted—is solved for σ_a and the plastic strain amplitude is substituted by a value of 0.002 corresponding to the 0.2% offset definition of the cyclic yield strength σ'_Y.

Since the HAZ properties are at the focus of the present considerations, only correlations based on hardness will be discussed in the following. An estimate for K′ has been proposed by Basan et al. (2010)

$$K' = 0.009(HB)^2 + 0.1173(HB) + 376.75 \tag{97}$$

and modified by Lopez and Fatemi (2012)

$$K' = \begin{cases} 9.8 \cdot 10^{-3}(HB)^2 - 1.26(HB) + 705 & for \quad R_m/\sigma_Y \leq 1.2 \\ 4.09(HB) + 613 & for \quad R_m/\sigma_Y > 1.2 \end{cases} \tag{98}$$

who also provide an equation for the cyclic yield strength

$$\sigma'_Y = 2.5 \cdot 10^{-3}(HB)^2 + 1.49(HB). \tag{99}$$

Equations (97)–(99) are based on Brinell hardness. For steels, a simple translation to Vickers hardness is given by

$$HV = 8.716 + 0.963(HB) + 0.0002(HB)^2 \tag{100}$$

JSMS (1998). Use can also be made of hardness conversion tables, e.g. ASTM E 140 (1997). Based on K' and σ'_Y, the cyclic strain hardening exponent n' can simply be determined by

$$n' = -0.37 \log\left(\sigma'_Y / K'\right) \tag{101}$$

Lopez and Fatemi (2012). Note that rather poor results are reported for estimates of n' based on the coefficients of the Manson-Coffin-Basquin approach (Roessle and Fatemi 2000).

Table 10 provides a comparison between experimental and estimated values for K' and n'. It distinguishes between two options. Equations (99) and (100) are combined with Eq. (97) in *Option A*, and with Eq. (98) in *Option B*. Note that only the mean values of the experimentally determined hardness have been used, which makes the comparison a bit uncertain. In Fig. 83 the same two options are applied to the stress amplitudes σ_a corresponding to a strain amplitude of $\varepsilon_a = 1.5\%$ when the estimates of K' and n' are combined by Eq. (96). Whilst Option A sometimes under- and sometimes overestimates this value, *Option B* yields an overestimation of σ_a in almost any case (except for the two base metals). Since conservatism in terms of the cyclic crack driving force requires underestimation rather than overestimation this seems to speak rather in favor of *Option A* than of *Option B*. However, keep in mind the uncertainties mentioned above because of which no final recommendation will be provided here.

2.3.2.4 Long Crack Propagation Curve (da/dN–ΔK Curve)

2.3.2.4.1 Curve Fit and Crack Closure Correction

The *long* crack da/dN–ΔK curve is fitted by Eq. (80), Sect. 2.2.4.1, with the crack closure factor U being U_{LC} in Eq. (81). This is determined by the (modified) NASGRO

Table 10 Comparison between experimental cyclic stress-strain curve parameters and those obtained by Eqs. (97 +99 +100 =*Option A*) and (98 +99 +100 =*Option B*); weld areas such as in Fig. 23

		K′ in MPa			n′		
Material	HV1	Exp.	*Option A*	*Option B*	Exp.	*Option A*	*Option B*
S355NL base metal	165	1203	641	1288	0.20	0.11	0.23
S355NL HAZ CG CP1	285	1527	1141	1779	0.17	0.10	0.17
S355NL HAZ CG CP2	330	1896	1396	1963	0.16	0.10	0.15
S355NL HAZ FG CP1	240	1539	923	1595	0.20	0.10	0.19
S355NL HAZ FG CP2	285	1744	1141	1779	0.19	0.10	0.17
S960QL base metal	330	1367	1396	1356	0.09	0.10	0.09
S960QL HAZ CG CP1	390	1729	1791	2208	0.10	0.10	0.13
S960QL HAZ CG CP2	420	1838	2014	2331	0.10	0.10	0.13
S960QL HAZ FG CP1	315	1520	1307	1901	0.13	0.10	0.16
S960QL HAZ FG CP2	360	1295	1585	2085	0.07	0.10	0.14

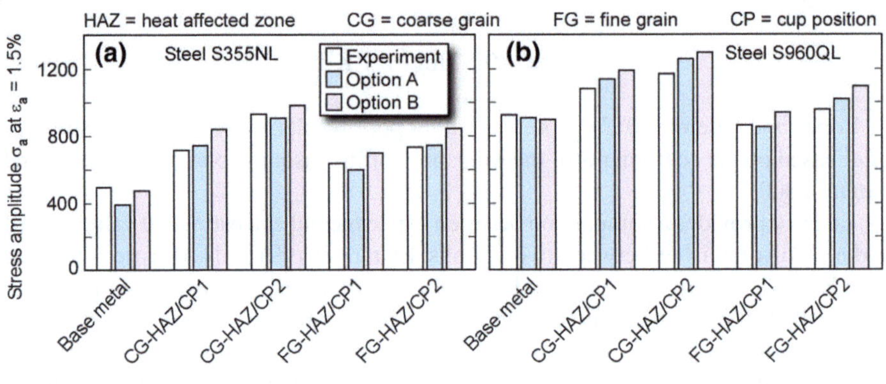

Fig. 83 Stress amplitudes obtained in the experiment and estimated using options A (Eqs. 97 + 99 + 100) and B (Eqs. 98 + 99 + 100) at the applied strain amplitude of 1.5% as well as deviations from the experimental values; **a** steel S355NL; **b** steel S960QL, IBESS data; according to Kucharczyk et al. (2018)

approach, Eqs. (82)–(88), see Sect. 2.2.4.2. For the IBESS basic procedure, the constraint parameter α_g is obtained as a fit parameter adapted to da/dN–ΔK curves at different R ratios, for explanation see Sect. 2.2.4.3. If no adequate experimental da/dN–ΔK curves are available, a value of $\alpha_g = 2.5$ can be used for steel at a lower analysis level in accordance with NASGRO (2000).

The fit of the threshold region of the da/dN–ΔK curve is provided by the second term of Eq. (80), $1 - (\Delta K_{th}/\Delta K_p)^p$. It is based on the fatigue crack propagation threshold ΔK_{th} and on the exponent p chosen such that it starts at $\Delta K_{th,eff}$ and then passes into the da/dN–ΔK_{eff} curve of the Paris regime.

2.3.2.4.2 Statistical Aspects

The discussion in this section is limited to the coefficient C of the Paris law (Eqs. 2, 3 and 4). Notes on the scatter of the long crack propagation threshold $\Delta K_{th,LC}$ will be provided in Sect. 2.3.2.5.2. For stochastic consideration in IBESS analyses, C should be available as a statistical distribution. A convenient measure for its characterization and for an immediate impression of the width of the scatter band is the coefficient of variation COV, i.e. the ratio of the standard distribution σ and the mean or expected value μ based on a certain type of mathematical distribution. Including BS 7910 (2013), statistical information is given in a number of documents and scientific papers. An extended overview for steel is provided by Josi (2010), a review of further data sets by Walbridge (2005). The summarized COV values of C refer to air and corrosive environment, low and high R ratios and specific application fields such as bridges or offshore structures. They are usually based on a lognormal distribution although data for normal distributions are published as well. The COVs range from 0.25 to 0.63 with the BS 7910 data at the upper level. There is, however some indication, see, e.g., Wu and Ni (2007) that these values might be too high and also that there might be some R ratio dependency of the scatter or, in other words, an effect of the crack closure phenomenon. This is confirmed by ongoing investigations of the present authors. The discrepancies are the reason why C has not been treated in a stochastic way in IBESS by now.

Instead, some basic aspects will be discussed in the following. It is important to distinguish between the real and an apparent scatter around the da/dN–ΔK curve. The principle is illustrated in Fig. 84, see also Sect. 5.1. of the USAF Damage Tolerance Design Handbook (Miedlar et al. 2002). When a da/dN–ΔK curve is obtained from a number of a-N curves of individual specimens (Fig. 84a) a scatter band can be identified by eye or statistical processing (Fig. 84b). Using the lower and upper bounds of this, the scatter band around the original a-N curves can be reproduced (Fig. 84c). However, what is usually found is that the confidence limits determined this way are significantly wider than those obtained for the original a-N curves. This prompts the conclusion that the scatter band in Fig. 84b is an apparent one. Various reasons might be responsible for this such as (i) successive phases of acceleration and deceleration of crack propagation, e.g., due to material inhomogeneity which are averaged out in the overall lifetime and, additionally, (ii) different techniques for smoothing the da/dN data before the analysis (frequency in data recording, secant or incremental polynomial method in data processing).

The conclusion is that the statistical distribution of C cannot be determined from the scatter in the data points in Fig. 84b. Instead a C value has to be determined for each individual a-N curve in Fig. 84a usually with a fixed value of the exponent n in Eqs. (2), (3), (4) and (80). The C distribution has then to be based on these data. The determination of replicate curves is time consuming and only a few data sets exist. The problem becomes even more complex if, as the authors suspect, crack closure will also have an influence on the scatter.

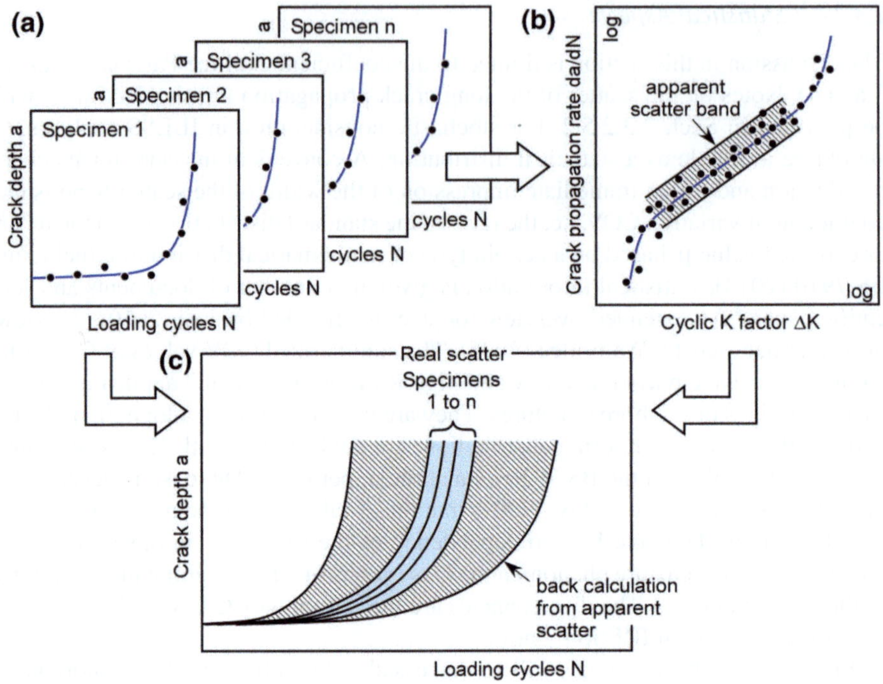

Fig. 84 Scatter and apparent scatter in fatigue crack propagation analyses, cf. Sect. 5.1 of Miedlar et al. (2002)

2.3.2.4.3 Reference Curves

At a lower analysis level, IBESS refers to the upper bound reference curves for da/dN–ΔK provided by BS 7910 (2013). For steels tested at air, the mean plus two standard deviations curve for $R \geq 0.5$ is given by

$$C = \begin{cases} 2.10 \times 10^{-17} \text{ for } \Delta K \leq 144\,\text{N/mm}^{3/2} \\ 1.29 \times 10^{-12} \text{ for } \Delta K > 144\,\text{N/mm}^{3/2} \end{cases}$$

and

$$n = \begin{cases} 5.10 \text{ for } \Delta K \leq 144\,\text{N/mm}^{3/2} \\ 2.88 \text{ for } \Delta K > 144\,\text{N/mm}^{3/2} \end{cases}$$

for da/dN in mm/cycle and ΔK in $\text{N/mm}^{3/2}$. Other reference curves such as the IIW curve (Hobbacher 2016) or the curve of Ohta et al. (1997) may be used as well. As can be seen in Fig. 85, they all provided meaningful upper bounds to the da/dN–ΔK curves of the two steels investigated within IBESS, for more details, see Kucharczyk et al. (2018).

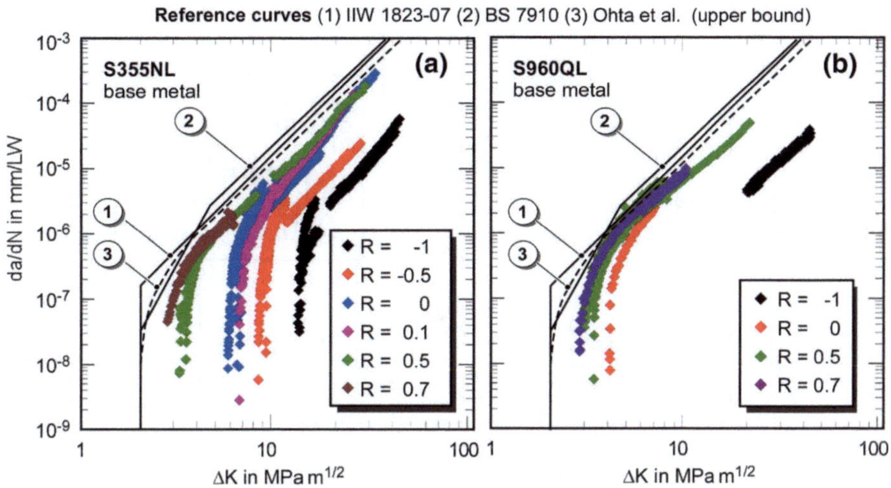

Fig. 85 Crack propagation curves da/dN–ΔK in comparison with reference curves; **a** steel S355NL; **b** steel S960QL, IBESS data; according to Kucharczyk et al. (2018)

2.3.2.5 Long Fatigue Crack Propagation Threshold ΔK_{th}

2.3.2.5.1 Experimental Determination

Not much has to be added here about the experimental determination of the *long* fatigue crack propagation threshold ΔK_{th} since this topic has extensively been discussed in Sect. 1.4.2 and in even more detail by Zerbst et al. (2016b). Caution is advised with respect to the conventional load reduction technique for materials which might be susceptible to corrosion. It is recommended to determine the parameter either (i) by compression pre-cracking testing or (ii) by load reduction testing at a high R ratio or (iii) by K_{max} testing, such as explained in Sect. 1.4.2.

Figure 86 shows a comparison between ΔK_{th} values obtained by conventional load decrease and compression pre-cracking tests for base metal and HAZ of S355NL steel. As can be seen, compression pre-cracking tended to lower thresholds, particularly for the HAZ material which was found to be susceptible to corrosion, i.e. a dark corrosion layer was found at the crack faces when tested at low R ratios.

2.3.2.5.2 Scatter

A topic on its own is the scatter in the fatigue crack propagation threshold. Particularly short cracks frequently show a huge scatter in ΔK_{th}. However, a closer look reveals that what is perceived as scatter at first sight, in reality is a zig-zag curve (Fig. 87) presumably caused by interaction effects between the crack and the local microstructure. Note that apparent scatter is a general problem in da/dN–ΔK testing (Sect. 2.3.2.4.2). Based on a range of literature data, Zerbst et al. (2016b) proposed a

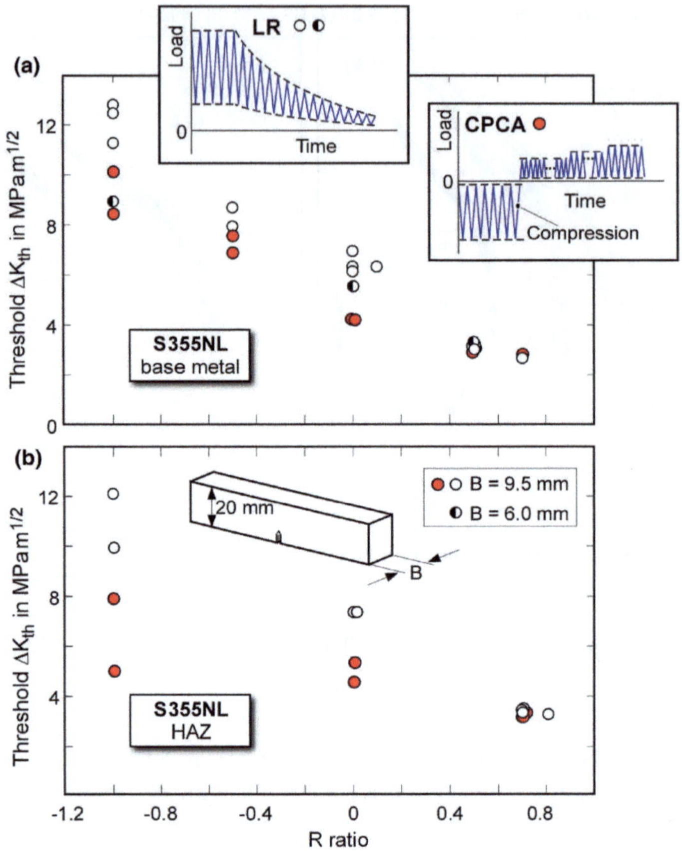

Fig. 86 Fatigue crack propagation threshold as a function of the R ratio determined by different methods; steel S355NL **a** base metal; **b** HAZ; IBESS data; according to Kucharczyk et al. (2018)

coefficient of variation for long crack thresholds of steel in the order of COV = 0.15 which, however, lacks further validation.

2.3.2.5.3 Reference Values

Reference values for the *long* fatigue crack propagation threshold are provided by BS 7910 (2013), among others, for welds. For steels tested at air or other nonaggressive environments and a temperature up to 100 °C a value of $\Delta K_{th} = 63\ N/mm^{3/2}$ (2 MPa m $^{1/2}$) is proposed, for aluminium alloys under the same conditions (but only up to 20 °C) a value of $\Delta K_{th} = 21\ N/mm^{3/2}$ (0.7 MPa m $^{1/2}$). Note that these values rather refer to the closure free intrinsic threshold. In addition, an R ratio dependent long crack threshold is provided for steels (excluding austenitic) in air:

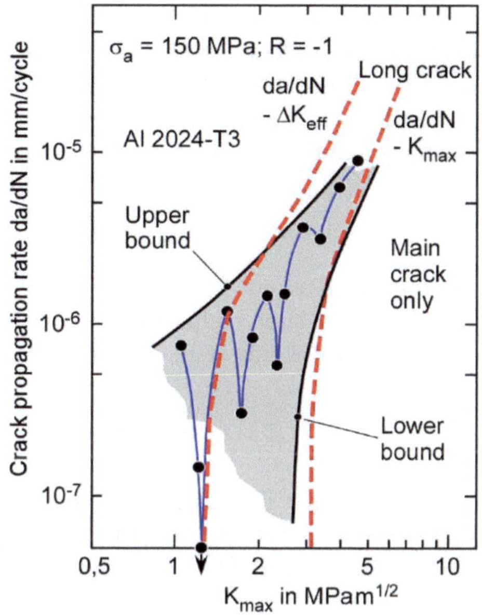

Fig. 87 Crack propagation for short cracks in 2024-T3 aluminium. Crack growth pattern; according to Tanaka and Akinawa (1989), data plotted according to McDowell (1996)

$$\Delta K_{th} \text{ in } N/mm^{3/2} = \begin{cases} 63 & \text{for } R \geq 0.5 \\ 170 - 214 \cdot R & \text{for } 0 \leq R \leq 0.5 \\ 170 & \text{for } R < 0 \end{cases}.$$

2.3.2.6 Short Fatigue Crack Propagation Threshold: Additional Information

The experimental determination of the cyclic R-curve has already been discussed in Sect. 1.4.3 and will not be repeated here. There is, however, one point to be added. The experimental curves should be approximated by an equation with at least two fit parameters. In IBESS a simple empirical equation

$$\Delta K_{th} = \begin{cases} A \cdot \Delta a^b + \Delta K_{th,eff} & \text{for } \Delta a < \Delta a_{LC} \\ \Delta K_{h,LC} & \text{for } \Delta a \geq \Delta a_{LC} \end{cases} \quad (102)$$

is used with A and b being these coefficients. It was found that one-parameter fits such as those provided by McEvily et al. (2003) are unsuitable for the description of the curve particularly in the initial part at low Δa (Zerbst and Madia 2015) which is highly relevant for later R-curve analyses; for a more detailed discussion see Maierhofer et al. (2018).

As well, the estimation method for the cyclic R-curve based on a modified El Haddad equation, Eqs. (6)–(8) has already been introduced in Sect. 1.4.3. The input information needed for the2.3.2.2 latter comprises (i) the material fatigue limit (range!)

which can also be estimated from parameters such as the ultimate tensile strength or the hardness, see Sect. , the *long* fatigue crack propagation threshold, see Sect. 2.3.2.5 and the intrinsic threshold.

The latter is a function of the elastic properties of the metal, i.e. the modulus of elasticity and the type of crystal lattice. A simple estimate is provided by

$$\Delta K_{th,eff} \approx 1.3 \cdot 10^{-5} E \tag{103}$$

(with E in MPa and $\Delta K_{th,eff}$ in MPa $m^{1/2}$), for a discussion see Maierhofer et al. (2018). An equation including the magnitude of Burgers' vector, $\|b\|$, was recently published by Pokluda et al. (2014):

$$\Delta K_{th,eff} = \frac{3}{4} E \cdot \sqrt{\|b\|} \tag{104}$$

Note that the $\|b\|$ value depends on the lattice type and is in the order of 0.25–0.29 nm for engineering metals such as steels or aluminium alloys.

The second key parameter for the determination of the cyclic R-curve by Eqs. (6)–(8) is the a_0 parameter proposed by Smith (1977) and El Haddad et al. (1979). A compendium of a_0 values is provided by Atzori et al. (2005) for a variety of materials. Note, however, that caution is advised when literature data are used. Besides the long crack threshold $\Delta K_{th,LC}$ and the material fatigue limit, a_0 also depends on the boundary correction function Y of the stress intensity factor (Eq. 6). Frequently this is assumed as $Y = 1$ without further consideration (obviously in order to provide it in a geometry independent format). Within the framework of IBESS, semi-circular or semi-elliptical surface cracks are considered, with the semi-circular ones being chosen at the fatigue stress level. For this, $Y = 0.728$ in tension loaded plates (Tanaka and Akinawa 2003). As a consequence, a_0 values referring to $Y = 1$ have to be converted to a_0 for $Y = 0.728$ when applied within the IBESS method. This is done by a "correction" factor of $(1/0.728)^2$. As already mentioned in Sect. 1.4.3, caution is required with respect to the long crack threshold $\Delta K_{th,LC}$ as an input parameter in Eq. (6) since a_0 is very sensitive to variations of this.

Note that the input parameters $\Delta K_{th,LC}$ and $\Delta\sigma_e$ in Eq. (6) depend on the load ratio R. With respect to the fatigue limit range $\Delta\sigma_e$ this means that, if necessary, a conversion to a different R ratio must be carried out such as is usual in fatigue. An option is provided by the Haigh approach, e.g., using Goodman's equation (Goodman 1914)

$$\sigma_a(R) = \sigma_e(R = -1) \Big/ \left(1 + \frac{\gamma(R) \cdot \sigma_e(R = -1)}{R_m}\right) \tag{105}$$

with $\gamma(R)$ being a conversion factor which can simply be determined as

$$\gamma = \frac{\sigma_m}{\sigma_a} = \frac{1 + R}{1 - R} \tag{106}$$

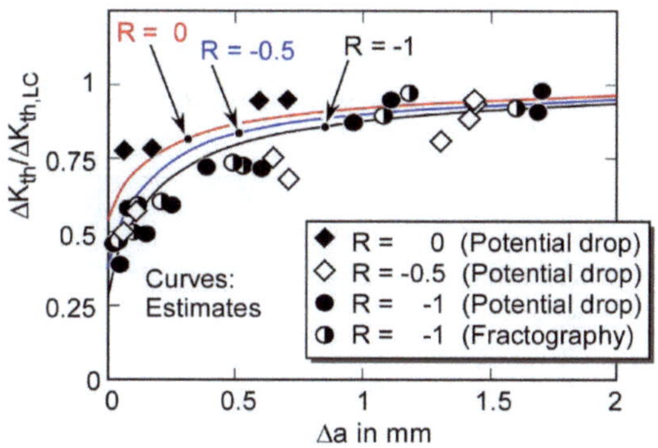

Fig. 88 Comparison of experimental cyclic R-curves of S355NL steel and estimates by Eq. (7); according to Zerbst and Madia (2015)

such that $\gamma = 0$ for $R = -1$, $\gamma = 1/3$ for $R = -0.5$, $\gamma = 1$ for $R = 0$ and $\gamma = 3$ for $R = +0.5$. Estimates of cyclic R-curves for different R ratios based on Eqs. (6)–(8) in conjunction with Eqs. (104) and (105) are provided in Fig. 88.

2.3.2.7 Monotonic Fracture Resistance

The topic of monotonic fracture resistance has already been addressed in Sect. 1.9.2.3. Note that, in IBESS, the data, besides those for the base metal, have been obtained on specimens with thermally simulated microstructure. Guidance on toughness testing on welded specimens is provided in test standards such as BS 7448, Part 2 (1997) and ISO 15653 (2010), see also Appendix 9 in EFAM GTP 02 (Schwalbe et al. 2002). With respect to the statistical treatment of weldment data, guidance is provided by Wallin et al. (2004), SINTAP (see, e.g., Sect. 2.4.5.3 in Zerbst et al. 2007, and Appendix A in Zerbst et al. 2014a), and in Sect. 7.1.5 and Annex L in BS 7910 (2013); cf. also Annex A of the present study.

2.3.2.8 Material and Weld Defects

Typical weld defects of relevance for fatigue strength are slag inclusions, microcracks at spatter droplets and other features, see e.g. Figure 16 in Sect. 1.5.1 and, for a more detailed discussion, Zerbst et al. (2014a). Note that no such flaws were found in the IBESS specimens. However, even if there are no defects from the welding process, flaws such as non-metallic inclusions exist in the base metal as well as in the weld metal from original material production. They are introduced into the material, e.g., during the deoxidation process, for a more detailed discussion see Zerbst et al. (2018b). Figure 89 shows the statistical size distribution (extreme value statistics) of such inclusions in S355NL welds, specifically in the weld toe regions. As explained in Sect. 1.6, the particle sizes of these inclusions cannot be simply used as initial crack sizes since it is not clear whether cracks initiated at them would arrest

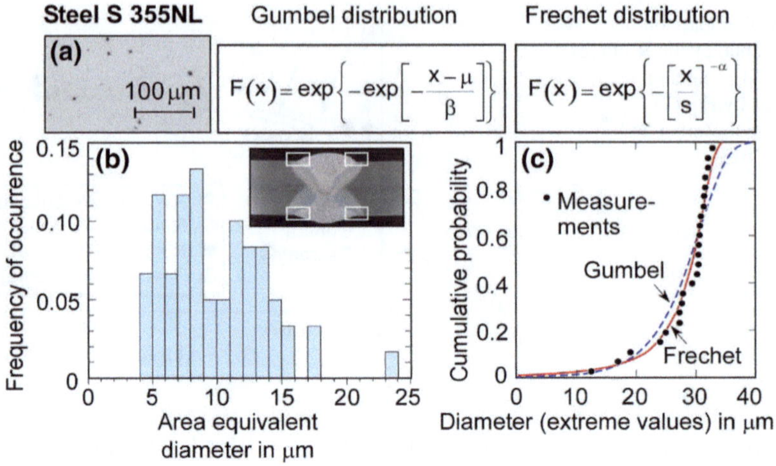

Fig. 89 Non-metallic inclusions in the weld toe areas of butt welds of S355NL steel. **a** Metallographic section; **b** histogram of the area equivalent diameter of a circle; **c** extreme value statistics of inclusion diameters; according to Schork et al. (2018)

later, such that they would not be critical for the component. They would only be relevant if their sizes exceeded the initial crack size at the transition from crack arrest to propagation. The determination of the latter within IBESS will be described in Sect. 2.4.5.

2.4 Determination of the Fatigue Strength and Life

2.4.1 Crack Arrest: Cyclic R-Curve Analysis

The cyclic R-curve analysis (the acronym "R" stands for the resistance of the material) is at the center of the IBESS procedure. The principle was first introduced by Tanaka and Akinawa (1988), for some more recent examples see McEvily et al. (2003), Chapetti et al. (2004) and Endo and McEvily (2007). It can be compared to the well-established R-curve analysis for monotonic fracture where the crack driving curve, e.g., in terms of the J-integral is plotted as a function of the crack size, a, along with the R-curve, J–Δa, with Δa being the amount of stable crack extension. In terms of the fracture mechanics triangle (see Sect. 1.5.1) the crack driving force stands for the "load vertex" and the R-curve for the "material vertex". The third, i.e., the "crack vertex" is defined by the (pre-existing) initial crack size for crack extension. Note that the crack driving force, besides the crack size, depends on the applied loading and the component geometry. The load, the crack driving force of which in combination with a certain initial crack size fulfills the tangency criterion with the R-curve, is that load at which the hitherto stable crack extension changes to unstable crack growth.

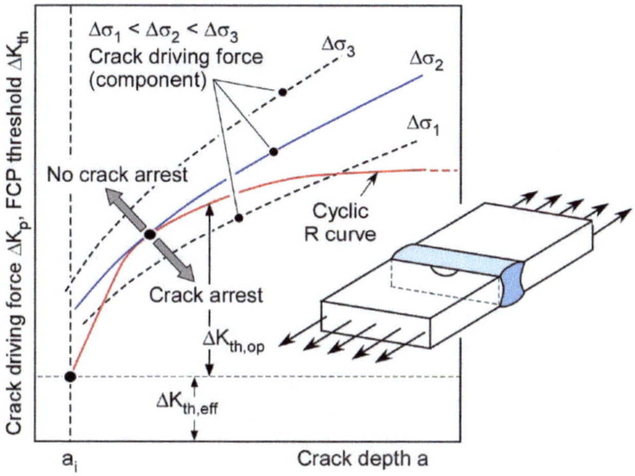

Fig. 90 Schematic view of a cyclic R-curve analysis. The transition from arrest to non-arrest is given by that crack driving force curve (referring to load $\Delta\sigma_2$ in the example) which tangentially touches the cyclic R-curve

Comparably, crack arrest turns to fatigue crack propagation when the tangency criterion is met in the cyclic R-curve analysis. In IBESS, the vertexes of the triangle are: (i) the crack driving force ΔK_p referring to the applied load $\Delta\sigma$ including local stress concentrators and the R ratio ("load vertex"), (ii) the cyclic R-curve $\Delta K_{th}(a)$ ("material vertex") and (iii) the initial crack size a_i and shape a_i/c_i of the semi-circular or semi-elliptical crack. The principle is illustrated in Fig. 90.

2.4.2 Statistical Description of the Weld Geometry Variation Along the Weld Toe

As already discussed, crack propagation from weld toes is a multiple crack problem influenced by the geometrical variation along the toe, see Fig. 20 in Sect. 1.8. This requires a specific kind of statistical treatment. As seen in Fig. 91, in the IBESS method the toe is "split" into a number of equidistant sections along its length. A set of geometrical parameters in terms of the toe radius ρ, the flank angle α, weld reinforcement h and secondary notch depths k is assigned to each section by "semi-randomly" sampling from the statistical distributions of these parameters. Note that a similar idea has been realized before by To et al. (1993) and Lecsek et al. (1995). Each section contains a half-circular initial surface crack the size ($a_i = c_i$) of which is also statistically distributed. For its determination, see Sect. 2.4.5.

The term "semi-random sampling" requires an explanation. It means that specific limits are introduced in random sampling to avoid too sharp geometrical transitions from section to section as they also would not exist in reality. Note that the parameters are chosen only within a certain confidence band, i.e., values with a cumulative

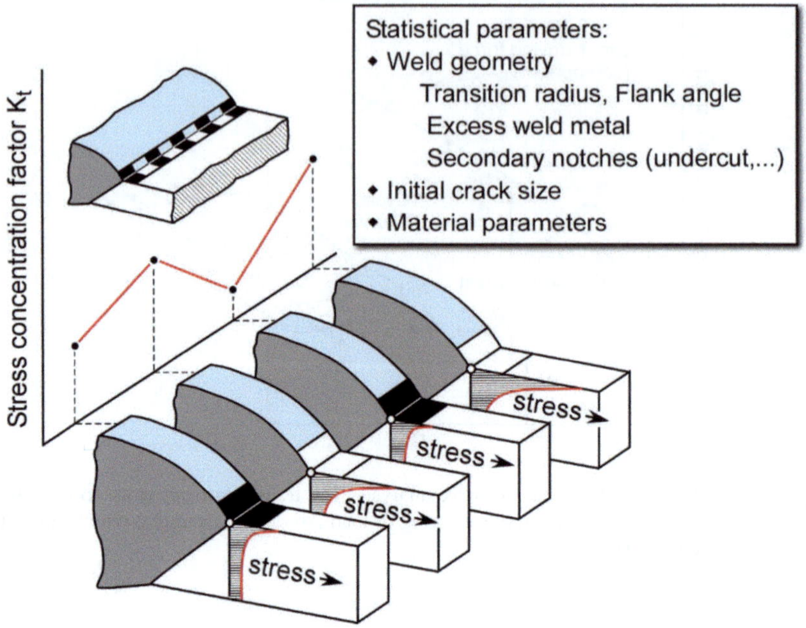

Fig. 91 Partitioning of the weld toe into equidistant strips, each of which is characterized by individual geometrical parameters generated by semi-random sampling from their statistical distributions; according to Madia et al. (2018)

probability of less than 10% and more than 90% are discarded. In addition, the random values for adjacent sections are chosen such that they alternate above and below the 50% value of cumulative probability, see Fig. 92. The sampling is performed by a Monte Carlo algorithm and the requirements above are simply fulfilled in that unwanted values are discarded before the analysis is continued.

A question that arises in this context is that of the appropriate choice of the width of the sections, in Fig. 92 designated by ℓ. Empirical evidence seems to point to the order of about 1–2 mm. Lecsek et al. (1995) chose 2 mm but did not find much a difference when they used 1.3 and 4 mm instead of this. In Fig. 93 surface roughness profiles along the weld toe (edge) and, parallel to this, at the center line of a S355NL butt weld are shown. For the first, some periodicity of the order of about 1.4 mm can be identified which is in line with the information above.

In order to evaluate the influence of the section width ℓ on the results of an IBESS analysis, calculations with $\ell = 0.33$–4.17 mm have been performed for a butt weld of S355NL at $R = -1$. The resulting mean curves are summarized in Fig. 94. As can be seen, calculations with $\ell = 1$ mm yielded most satisfying results. However, the difference within the complete range of ℓ values is no more than 25% in finite life (high cycle) fatigue strength (which roughly refers to the experimental scatter band) and even a bit less towards the fatigue limit. As the result of all these considerations, a tentative section width of $\ell = 1$ mm was chosen for the further analyses.

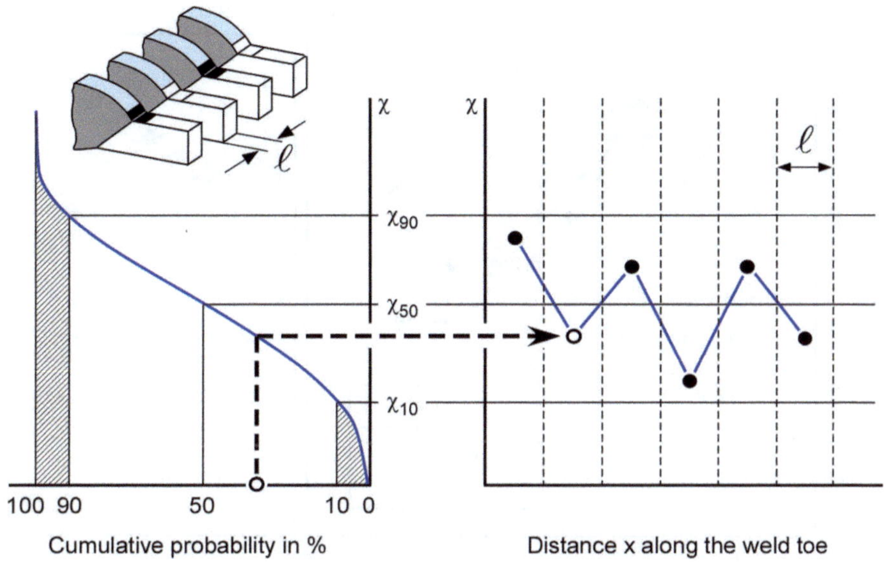

Fig. 92 Scheme of semi-random sampling of the geometrical parameters in order to avoid unrealistic sharp transitions between adjacent sections; according to Madia et al. (2018)

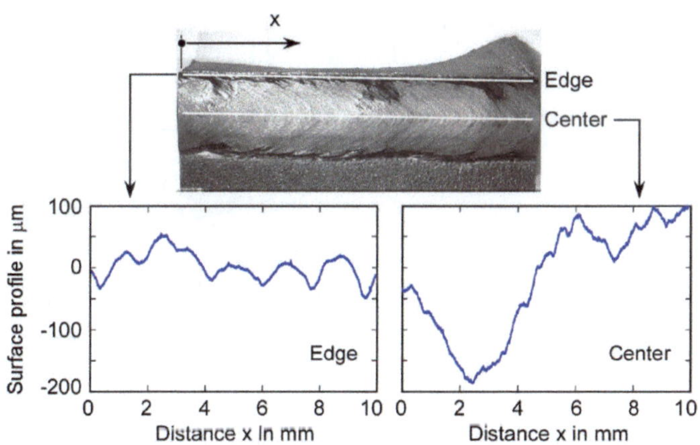

Fig. 93 Example of the ripple spacing determined by line scans at two straight lines along the weld toe and the center of the weld, material: S355NL, butt weld, IBESS data; according to Schork et al. (2018)

2.4.3 Crack Coalescence and Crack Aspect Ratio a/c

Within the IBESS method, each section contains an initial semi-circular surface crack the depth a_i of which is determined by the method described in Sect. 2.4.5.

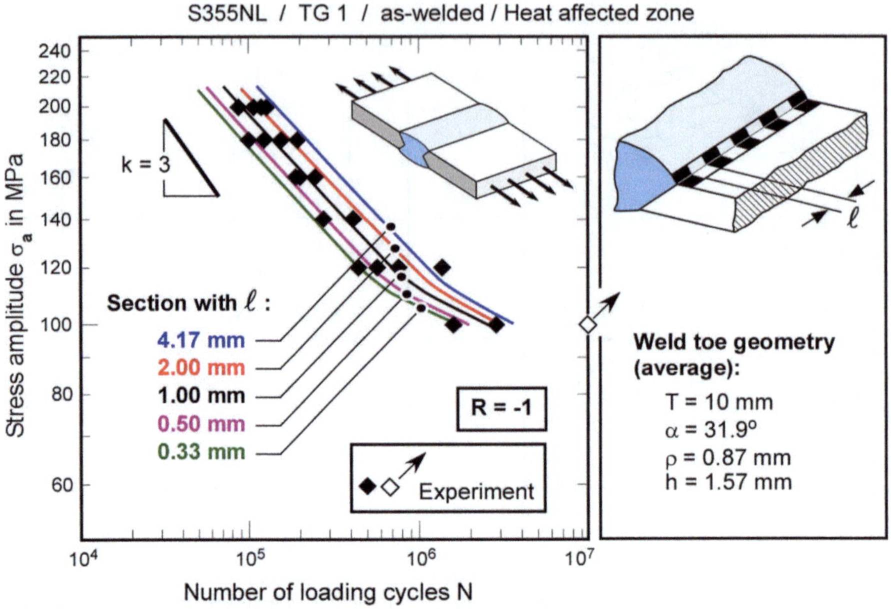

Fig. 94 Example for IBESS simulations of the S-N curve of a butt weld as a function of the section width ℓ along the weld toe, steel S355NL (HAZ); according to Schork et al. (2018)

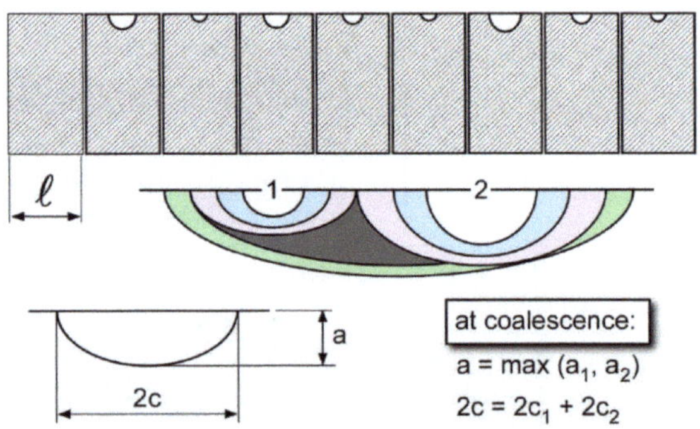

Fig. 95 Treatment of the coalescence of adjacent cracks in IBESS

The parameter shows a scatter band such that each section contains a crack of a different size. Depending on the different geometry of the weld toe from section to section some of these cracks will grow faster, others slower and some will even arrest after some propagation. A more detailed discussion of this pattern is provided by Madia et al. (2017a). When, subsequent to some crack extension, the surface points of two adjacent cracks contact, crack coalescence is assumed. The surface length of the resulting crack is equal to the combined surface lengths of the coalescing cracks and its depth is equal to the depth of the deeper crack. This principle is illustrated in Fig. 95.

Note that this rule is different from what is common for long cracks, see, e.g. Anderson (2003) and that it is a simplification in two respects: due to mutual influencing of the two crack tips at the adjacent surface points their crack driving force is larger than it would be if there were no second crack. As the consequence, coalescence at this point is faster than predicted by IBESS. In contrast, crack growth is slower immediately after this event when the propagation in the depth direction is decelerated whilst IBESS assumes the generation of the new semi-elliptical crack in a moment. Note that there is ongoing investigation by the authors of this pattern. By now, short crack coalescence is realized in IBESS as discussed above. The solution can be assumed as reasonably accurate or slightly conservative (Pang and Gray 1993; Tan and Chen 2013).

A further source of conservatism is the neglection of the "waviness" in the weld toe which is modelled as a straight line along its length dimension in IBESS. In reality, adjacent cracks are not strictly coplanar but show some offset of plane, which hampers and decelerates coalescence (Otegui et al. 1991).

A comparison between IBESS analysis results and experimental information is provided in Fig. 96. In the experiment, the crack development has been marked by heat tinting after 2.5×10^5 loading cycles (black and white areas respectively at the lower edge of the cross section in Figs. 96a, b) and beach marking (at later stages). The IBESS simulation of Fig. 97 does not start at zero but with the sizes of the cracks visualized by heat tinting. As can be seen, a quite similar picture emerges in experiment and simulation. The growth of one of the marked cracks of Fig. 96 is further pursued in Fig. 97 where satisfying analysis results can be attested for the crack depth (Fig. 97a) as well as for the crack length at surface (Fig. 97b). The zig-zag pattern of the crack aspect ratio a/c (Fig. 97c) is due to repeated coalescence events and it is in line with what is known from the literature, see Fig. 19 in Sect. 1.7.

2.4.4 Failure Criteria

In general, failure may be defined as "loss of functionality", see Sect. 1.1.5. This may occur as fracture but other criteria are possible as well. A common approach in fatigue analysis is an end-of-life definition by a certain crack depth relative to the wall thickness. This sounds arbitrary. Note, however, that not much lifetime is spent during the late phase of crack propagation such that the error is rather small. Although the monotonic fracture resistance has been determined in IBESS as well (Sect. 1.9.2.3), the validation exercises of Sect. 2.7 are based on the simple criterion that the depth of the crack has reached the order of 50% of the wall thickness. Note that the crack has shaped into a continuous surface crack at that point due to repeated coalescence events.

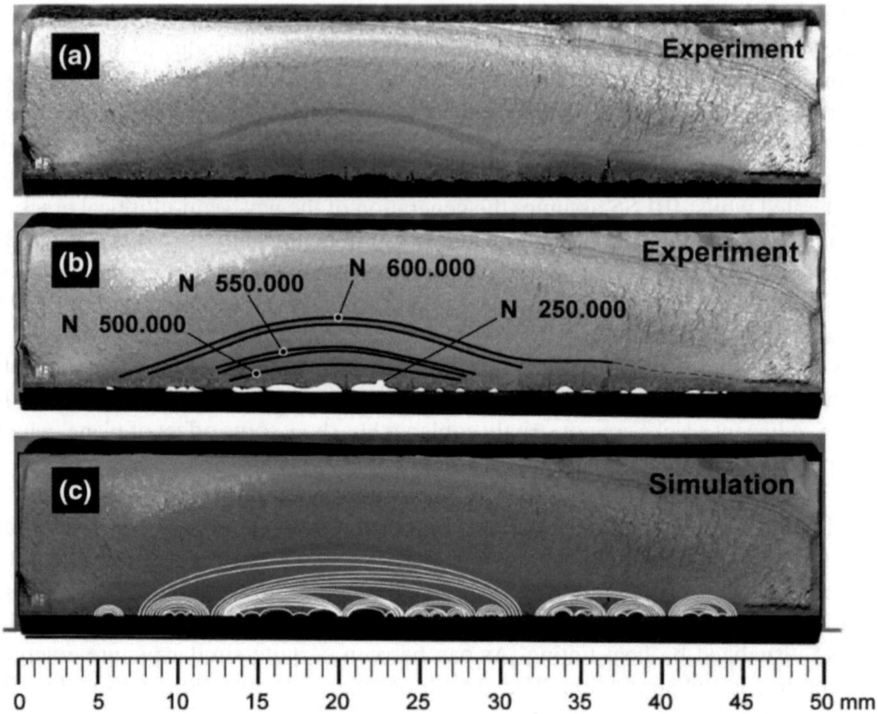

Fig. 96 Multiple fatigue crack propagation; **a** marked cracks (by heat tinting and beach marking); **b** replotted pattern with the corresponding numbers of loading cycles; **c** results of the IBESS simulation starting from the cracks visualized by heat tinting (black and white areas respectively at the lower edge in **a** and **b** as initial cracks; steel S355NL; R = −1; according to Madia et al. (2018)

2.4.5 Determination of the Initial Crack Size in IBESS

It has already been mentioned in Sect. 1.6 that the initial crack size a_i is determined by a two-criteria concept. The lower bound of a_i is obtained by a crack arrest analysis. If crack-like defects exist larger than the arrest crack size these take over the role as initial cracks. This is schematically illustrated in Fig. 98.

The determination of a_i as crack arrest depth is based on a cyclic R-curve analysis as described in Sect. 2.4.1. It is illustrated in Fig. 99. First, the crack driving force curve ΔK_p (a) is determined such as described in Sect. 2.2.3.2 for semi-circular surface cracks (a/c = 1) of different size. The load refers to the fatigue limit stress of the material which, if it is not explicitly available, can also be estimated from parameters such as the ultimate tensile strength R_m or the hardness HV, Sect. 2.2.3.2.

The second information available is the cyclic R-curve which is fixed at the ordinate of the diagram at the intrinsic fatigue crack propagation threshold $\Delta K_{th,eff}$ as its minimum value, but it is free to be shifted along the abscissa, i.e., the crack depth axis. The arrest-propagation transition is defined such that the crack driving force

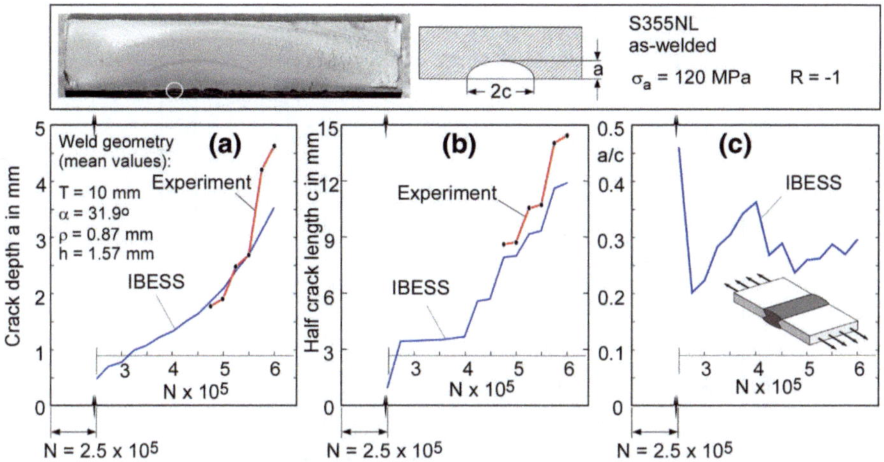

Fig. 97 Propagation of one of the cracks of Fig. 96; experimental results and IBESS analysis starting from the crack size marked by heat tinting; **a** development of crack depth, (a); **b** development of (half) crack length at surface, (c); **c** development of the crack aspect ratio, a/c; steel S355NL; R = −1; according to Madia et al. (2018)

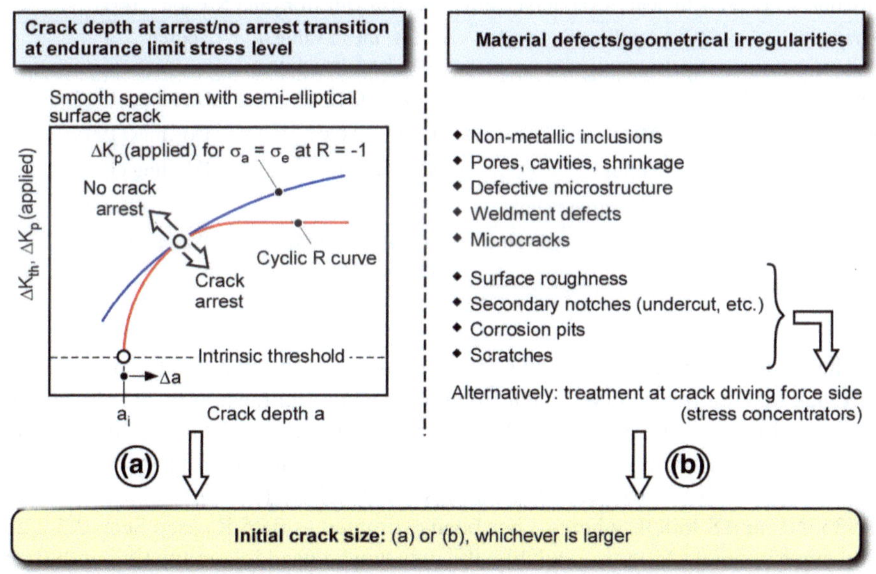

Fig. 98 Definition of the initial crack size for fracture mechanics analyses by either crack arrest or initial flaws, whichever provides larger values; according to Zerbst et al. (2017, 2018c, d)

curve and the R-curve just touch tangentially. This is reached by horizontally shifting the R-curve. The corresponding initial point of the cyclic R-curve at the abscissa refers to the initial crack size a_i.

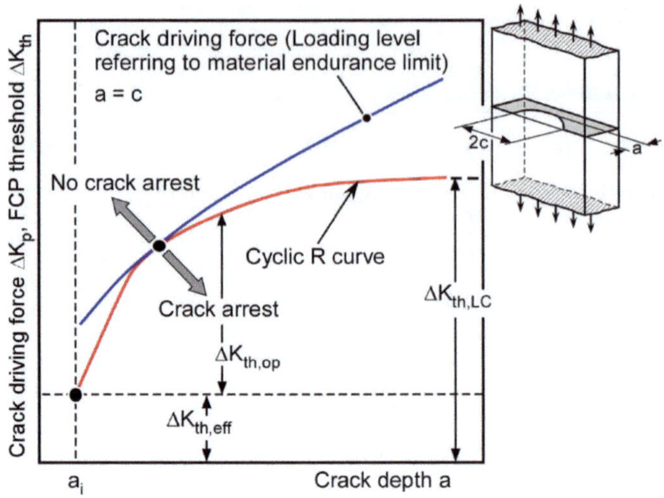

Fig. 99 Cyclic R-curve analysis for determining the initial crack size as a material parameter

Table 11 Reference data for probabilistic analyses (μ = mean or expectation value, σ = standard deviation, COV = coefficient of variation = σ/μ), see also Dillström and Nilsson (2003)

Parameter	Application range	COV, mean value μ standard deviation σ	Source
Fatigue strength		COV = 0.10	Dowling (1998)
R_m	Metals	COV = 0.05	Dowling (1998)
R_m	Welds	COV = 0.10	Dowling (1998)
R_m	Measured values	COV = 0.05	Wallin (1998)
R_m	Standardized values	$\mu = R_m + 70$ MPa $\sigma = 30$ MPa	Wallin (1998)

The value of a_i determined in this way is considered as a material parameter since the crack driving force curve is determined for a smooth tension plate subjected to cyclic loading at a stress ratio of $R = -1$. When the material fatigue limit (or the ultimate tensile strength or hardness) are available as statistical properties, a_i will also be statistically distributed. This is illustrated in Fig. 100 for the base materials of S355NL and S960QL where σ_e has been estimated as 0.45 R_m, see Sect. 2.3.2.2b. If no experimental scatter is available, the data can be used in conjunction with fixed coefficients of variation COV. Default values are provided in Table 11.

By the way, Fig. 100 can be used to explain the well-known phenomenon that the fatigue strength of high-strength steels is much more sensitive to the size of inclusions (Murakami 2002; see also Neuber 1958; Peterson 1959 for background explanation) than those of lower-strength steels. As can be seen, the initial crack size based on the arrest criterion of the higher strength steel S960QL is significantly smaller than those of the lower strength steel S355NL. Remember the two-parameter concept for

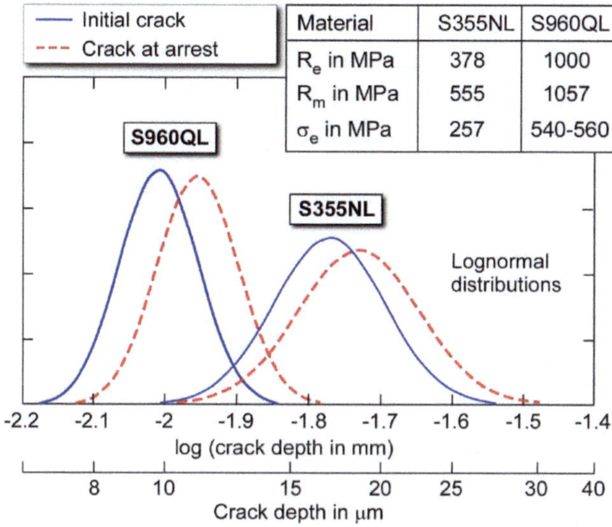

Fig. 100 Initial crack sizes and crack sizes at crack arrest obtained by the IBESS crack arrest model of Fig. 99 for low strength and high strength steels (here: base metals of S355NL and S960QL); according to Zerbst et al. (2018c, d)

defining the initial crack size (Fig. 98). Due to the lower crack arrest a_i in the higher strength material there is a higher probability for the largest inclusion to take over the role as the initial crack-like defect, see also Zerbst et al. (2018b, c, d).

2.4.6 Stochastic Determination of the Fatigue Strength of the Weldment

2.4.6.1 Basic Principle

If the initial crack size is known, the cyclic R-curve analysis can be applied to the determination of the fatigue limit of components, e.g. Akinawa et al. (1997). The basic principle is illustrated in Fig. 101 where each point stands for an IBESS analysis of a randomly sampled geometry. The stochastic input parameters, which differ from section to section of the weld toe, are the weld toe radius ρ, the flank angle α, the weld reinforcement h, the secondary notch depth k and the depth a_i of the initial semi-circular crack such as described in Sect. 2.4.5.

(a) At the finite life (high cycle) regime of the S-N curve, structural failure (for the definition see Sect. 2.4.4) will be reached at a number of loading cycles which differs from random geometry to random geometry. The result is a scatter band of failure loading cycles per stress level. Usually sets of such scatter bands will be generated at stress levels between the fatigue limit and the cyclic yield stress level which, finally, can be statistically processed.

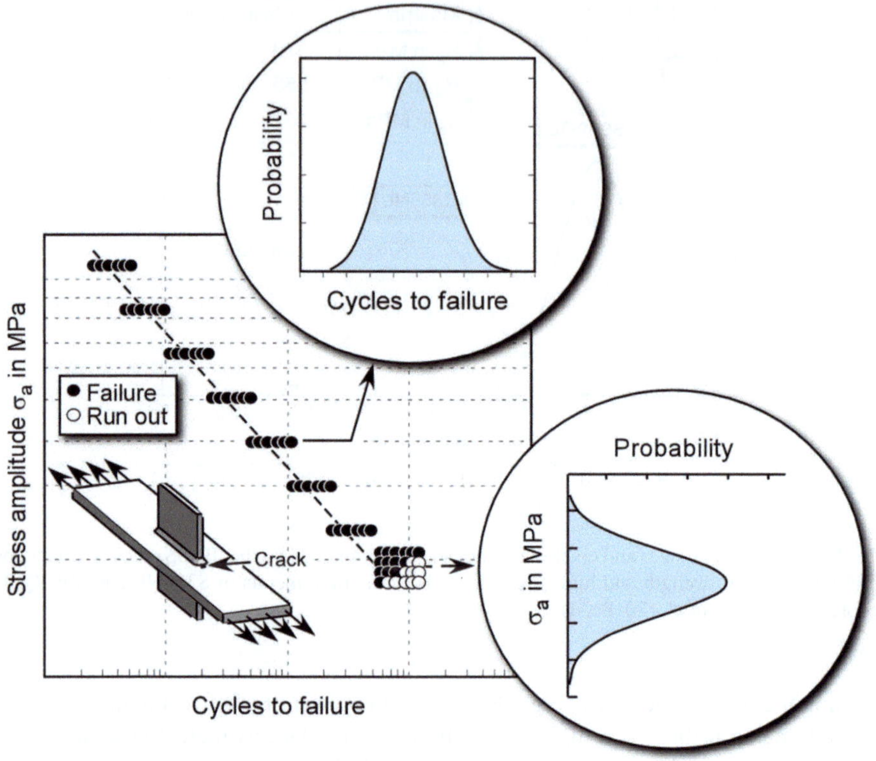

Fig. 101 Basic principle of the stochastic analyses of IBESS for the determination of the finite life (high cycle) S-N curve and the fatigue limit

(b) With respect to the endurance limit, an engineering definition is applied such that no failure will occur within 10^7 loading cycles (a number which, in principle, can freely be chosen). A pre-defined number of analyses is performed at different stress levels. At each of those levels, a number of failure events (before reaching the 10^7 loading cycles) and a number of runouts (no failure even at 10^7 loading cycles) is predicted. The failure probability is simply obtained as the ratio of failure events to the overall number of simulations. It will be zero at low stress levels and then increase until it reaches an upper bound fatigue limit above which the failure probability will be 100%. Based on these data, a probability function of the fatigue limit can be determined.

2.4.6.2 Fatigue Limit and Finite Life (High Cycle) Fatigue Strength

This principle is explained a bit more in detail in Fig. 102. An example is provided in Fig. 103 where the analysis is performed for both base metal and HAZ material

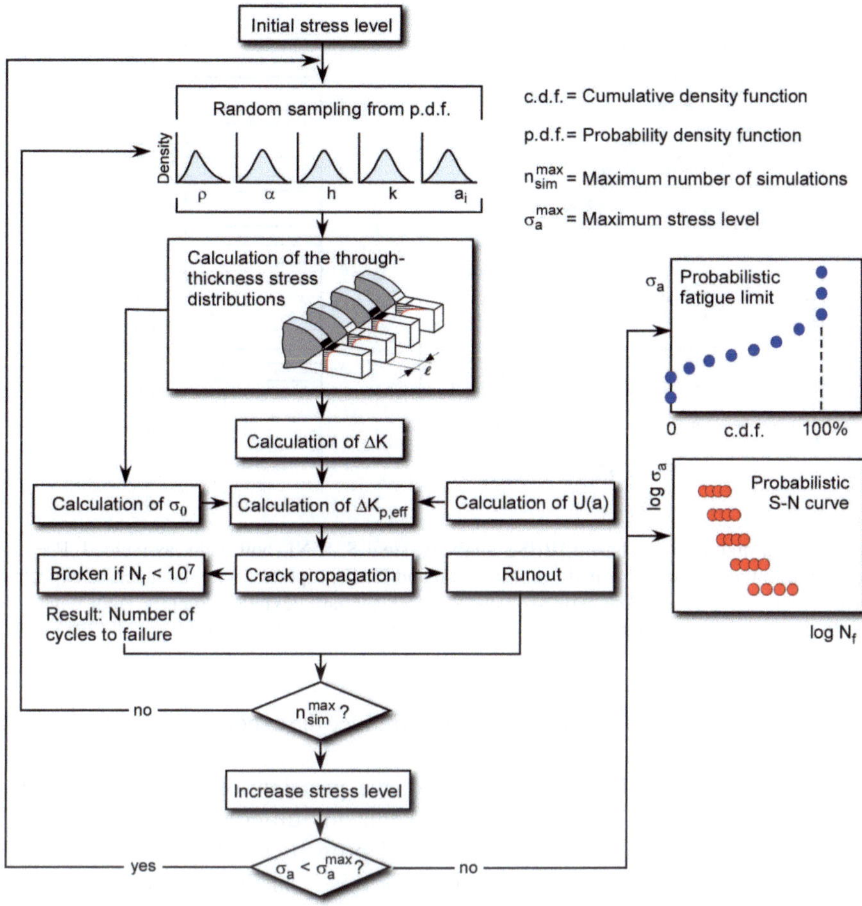

Fig. 102 Stochastic analysis of IBESS for determining the probability function of the fatigue limit with the latter being defined for $N = 10^7$ loading cycles and the scatter band of the finite life (high cycle) S-N curve

properties. As expected (see Sect. 1.9.2.1) the HAZ analysis provides the better results compared to the experimental data. Further validation examples will be provided in Sect. 2.7.

2.4.7 Proposal for a Simplified Procedure Based on Base Metal Data

The determination of the HAZ material data (which are needed because the short crack spends most of its time in that material region) is a sophisticated and time-consuming task. Therefore, options for the estimation of those data have been pro-

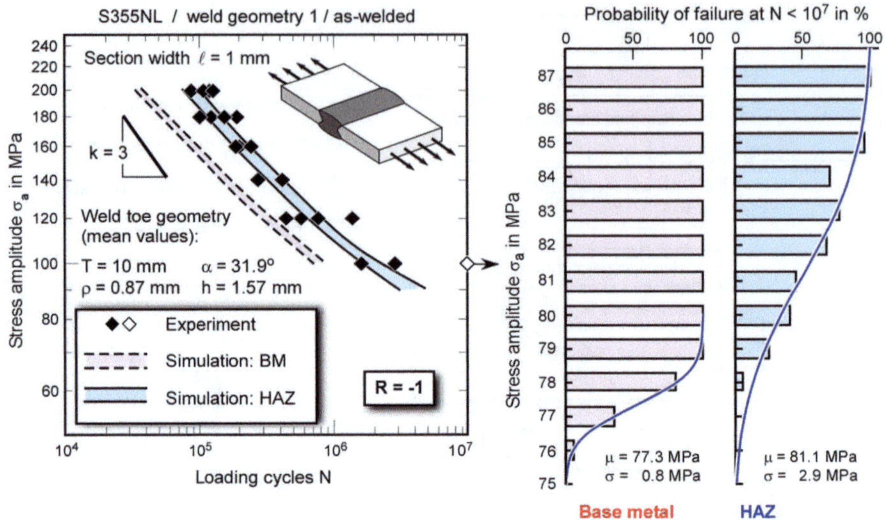

Fig. 103 Example of a stochastic IBESS analysis; steel S355NL butt weld, as-welded, $R = -1$. The simulation has been performed for base metal and HAZ properties with the latter, as expected, delivering the better result. The expected values μ and standard deviations σ at the right-hand side refer to the fatigue limit defined for $N = 10^7$ loading cycles; according to Madia et al. (2018)

posed, e.g. on the basis of hardness, in order to reduce the expense, see Sects. 2.3.2.2b and 2.3.2.3b.

Here, another simplification is proposed, the principle of which is illustrated in Fig. 104. It is based on the observation that the difference between the S-N predictions for base metal and HAZ material properties reduces at lower stress levels towards the fatigue limit. The reason for this is easily explained. The major differences between the base metal and HAZ cyclic stress strain curves are in the non-linear parts of the S-N curves whilst almost no difference is seen at the elastic curve sections, see Fig. 27 in Sect. 1.9.2.2. At the fatigue limit, most of the differences is given by different $\Delta K_{th}(a)$ characteristics (ΔK_{th} (HAZ) > (ΔK_{th} (BM)).

The simplified method consists of the following steps:

(a) The finite life (high cycle fatigue) S-N curve is stochastically simulated such as described in Sect. 2.4.6, but for base metal properties.
(b) To its upper tail a lower bound curve is plotted as a straight line in double-logarithmic scaling.
(c) This straight line is extrapolated to $N = 2 \times 10^6$ loading cycles (black dot in the figure).
(d) Then a line with a slope of $k = 3$ is plotted through this point with k being defined by

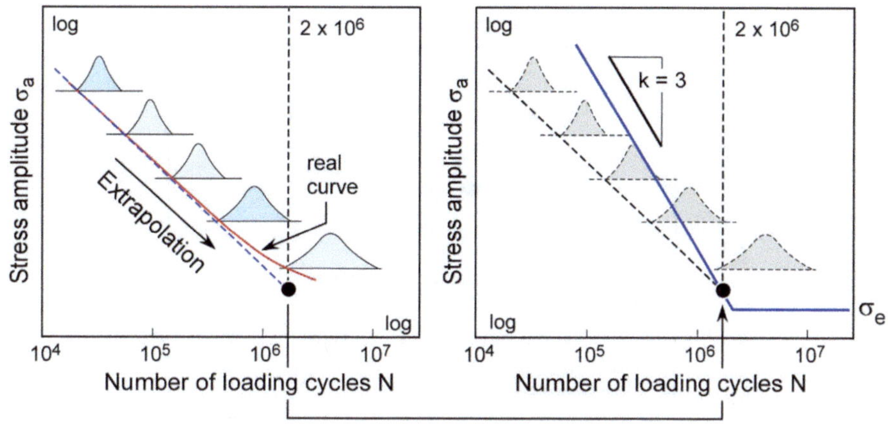

Fig. 104 Proposal for a simplified IBESS analysis based on base metal data. Based on the upper tail of the simulated S-N curve for base metal, a stress amplitude is determined for $N = 2 \times 10^6$ (left-hand side). A "synthetic" finite life branch with a slope of, e.g., $k = 3$ is then plotted through this point (right-hand side). Finally, the knee point is given by the intersection between a lower bound endurance limit (horizontal) and this line

$$k = \Delta(\log N) / \Delta(\log \sigma_a). \tag{107}$$

Note that a slope of $k = 3$ is given for S-N curves of weldments, see, e.g. the IIW guidelines (Hobbacher 2016).

(e) The fatigue limit is determined such as described in Sect. 2.4.6 and a horizontal line is plotted though its lower bound. This can be statistically defined as a lower percentile value or simply placed by eye when the scatter in the fatigue limit is small which is usually the case for larger flank angles α whilst the scatter band becomes wider for small α.

(f) The knee point is simply determined as the intersection point between the two straight lines in double-logarithmic scaling.

An example of a S-N curve determined this way is shown in Fig. 105.

2.5 Preliminary Recommendations for Considering Welding Residual Stresses

The treatment of residual stresses has been discussed in Sect. 1.9.4.3. Because of a number of open points the recommendations in this Section are preliminary.

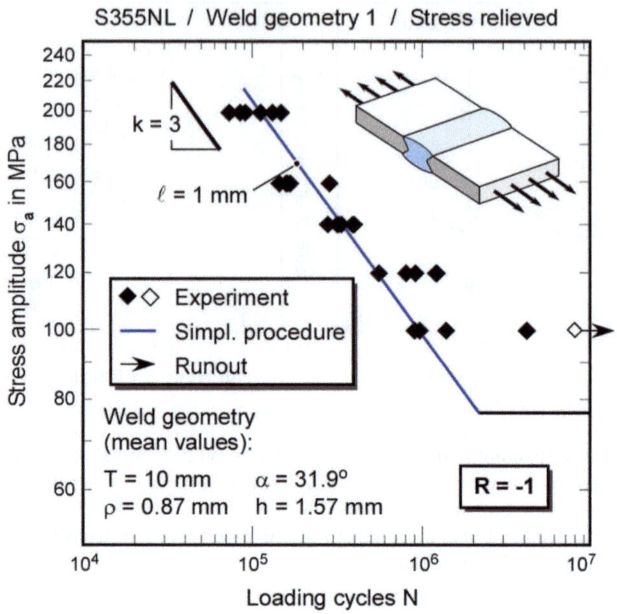

Fig. 105 IBESS simplified analysis based on base metal data, example; steels S355NL, butt weld, R = −1

(a) **Classification as Primary Stresses**

As a rule, residual stresses may be classified as primary in IBESS. This is particularly indicated when

- substantial elastic follow-up is expected,
- load controlled situations are expected, e.g., when the locality of the weld is far from the source of the displacements, (which is typical for reaction stresses) and
- in the case of *short* cracks when the spatial extend of residual stresses is large compared with the crack size.

The last two points are in accordance with BS 7910 (2013). Note that the treatment of welding residual stresses as primary in IBESS fatigue crack propagation analysis is not expected to be overly conservative since the *long* crack stage particularly in *Region* 3 of the da/dN–ΔK curve (see Fig. 5 in Sect. 1.4.1) and the final fracture event are usually not very relevant with respect to the overall lifetime. That is also the reason why *Region* 3 is not considered in the IBESS description of crack propagation (Eq. 80, Sect. 2.2.4.1).

(b) **Effective R Ratio**

The effect of welding residual stresses is considered by a modified R ratio according to Eq. (15) in Sect. 1.9.4.4. It is introduced in the analysis on the material side, namely in Eqs. (81) and (82). As outlined in Sect. 1.9.4.4, this R is understood as an effective

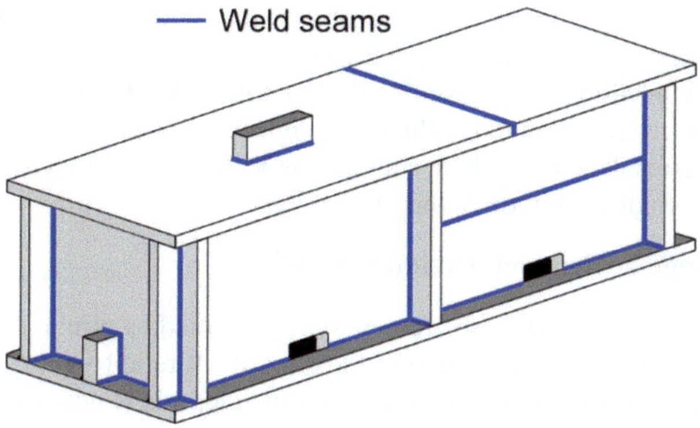

Fig. 106 Constructed example of a welded component with high internal restraint

value which replaces the nominal R exclusively defined for applied loading. The objection that this is not adequate when crack closure comes into play is irrelevant at this reflection because crack closure is adequately considered by Eq. (81).

(c) **No Residual Stresses at the Weld Toe in Not Self-restrained Weldments**

In the absence of *long range* (or reaction) residual stresses the remaining *medium-range* stresses are assumed to be zero in an IBESS fatigue analysis. The distinction between both kinds of residual stresses and its separation is, however, possible only in the absence of self-restraint of the weldment. It is possible for butt welds, T joints and cruciform joints but not for longitudinal gussets and, of course, it is also not the case for complex weldments such as shown in Fig. 106 as an example. There are two reasons for the zero-stress assumption (see Sect. 1.9.4):

- The as-welded residual stresses tend to values of zero at the weld toe region and, even if not,
- they are relieved by fatigue loading. Stress relief effects are also expected in monotonic load cases such as proof tests, local compression etc.

However, some caution is advised. Have a look at Fig. 38 in Sect. 1.9.4.3.1. Whilst the transverse residual stresses of the butt welds are zero at the weld toe surfaces, there are tensile residual stresses in the wall thickness direction. These are as-welded residual stresses and we do not know how they would develop under cyclic loading. If they were not completely removed, some (more or less moderate) non-conservatism of the IBESS analysis would be the consequence.

What also can be seen in Fig. 38 is a difference between the as-welded stress profiles referring to MAG and TIG welding. Note that the difference between the techniques consists in different heat inputs and $t_{8/5}$ temperatures as a measure of the cooling rate ($t_{8/5}$ is the time for cooling of a welding layer from 800 to 500 °C). These, in conjunction with the temperature-transition-time (TTT) diagram of the material will not only affect the microstructure but also the temperature of phase transformation from fcc to bcc lattice which in turn will alter the residual stress state.

No benefit should be taken from potential compressive residual stresses when, e.g., the weldment is cycled at positive R ratios.

No recommendation can be given at present with respect to post-weld treatment by the various peening methods which also introduce local compressive residual stresses. The crucial point is their stability under cyclic loading which will also depend on the strain amplitude of the fatigue load.

(d) **Long Range or Reaction Residual Stresses**

As outlined under point (a) these are generally classified as primary and they have to be considered in the determination of the effective R ratio (point b). What makes the issue complex is their determination. In principle, this is possible by finite elements (and in a few cases even analytically). However, their magnitude will strongly depend on manufacturing tolerances and it is definitely not an easy task to cope with these and other factors influencing the reaction stresses (see Sect. 1.9.4.1b), e.g. in a complex steel framework. If no or no reliable information can be gained the general treatment of the problem such as proposed by BS 7910 (2013) is recommended, i.e., individually determined or reference da/dN–ΔK curves (Sect. 2.3.2.4.3) for R $\geq$ 0.5 should be used. Note that Eurocode EN 1993-1-10 for reaction stresses implicitly assumes a default value of 100 MPa for steel structures (Sedlacek et al. 2008).

What is not recommended for the reasons discussed in conjunction with Fig. 38 in Sect. 1.9.4.3.1 is the use of compendia of residual stress profiles, e.g. of BS 7910 (2013) in the context of an IBESS analysis.

2.6 *Misalignment*

Figure 107 illustrates the two basic types of misalignment:

- Axial misalignment (Fig. 107a) and
- Angular misalignment (Fig. 107b).

Both can be the result of poor manufacturing but also of detailed design and they cause misalignment stresses when the weldment is subjected to membrane (tensile)

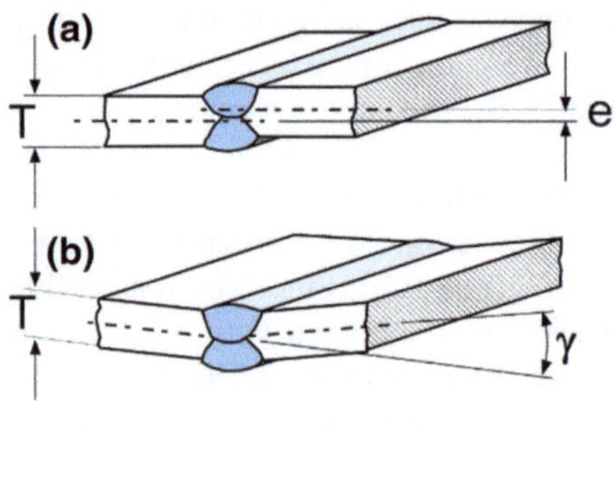

Fig. 107 Definitions of misalignment; **a** axial; **b** angular

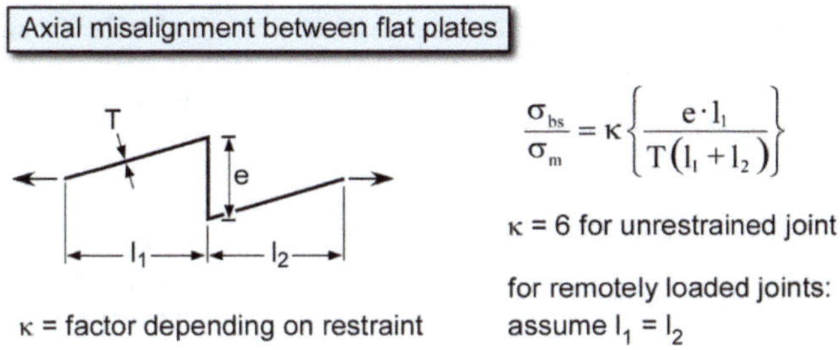

$$\frac{\sigma_{bs}}{\sigma_m} = \kappa\left\{\frac{e \cdot l_1}{T(l_1 + l_2)}\right\}$$

Fig. 108 Determination of bending stresses σ_{bs} due to axial misalignment in BS 7910 (2013); example

loading. No additional stresses arise in butt and fillet welded joints subjected to bending, however, this pattern might be different with respect to misaligned joints in sections, e.g., beams or tubes (BS 7910, 2013). In IBESS, misalignment is considered as an additional bending stress added by superposition in K-factor determination (Sect. 2.2.3.2.5) and as a reduction of the tensile reference yield load, in the determination of the reference yield load, σ_0 (Sect. 2.2.3.2.6, Fig. 56).

The determination of the misalignment stresses is not an objective of the procedure but the information has to be provided as an input information. Note that this usually will require a finite element analysis since the misalignment stress besides eccentricity, e, (axial misalignment) and angle, γ, (see Fig. 107) depends on further factors such as the section shape, the presence of other members in the welded assembly which provide local stiffening, etc. (BS 7910, 2013).

For basic cases, documents such as BS 7910 (2013) or the IIW guidelines (Hobbacher 2016) provide estimates of the additional bending stress referred to the applied membrane stress, σ_{bs}/σ_m. An example is shown in Fig. 108.

2.7 Application and Validation of the Approach

2.7.1 Case Studies—Overview

(a) Weldments Under Investigation

The method was applied to a large number of welds comprising

- different weldment types: butt, cruciform, both sided longitudinal gusset (Fig. 109),
- different materials and plate thicknesses (S355NL and S960QL; T = 10 mm; S355J2+N: T = 3 mm) with
- two weld geometries obtained by different manufacturing techniques, MAG and TIG respectively (Figs. 110 and 111), and in the as-welded and stress relieved state.

Table 12 Fit parameters A and B (Eq. 102) to the cyclic R-curves used in IBESS analyses; according to Kucharczyk et al. (2018)

Material	R ratio	A	b	Remarks
S355NL base metal	−1	6.292	0.412	
	0	1.924	0.299	
	0.5	0.37	0.5	Estimated
S355NL CGHAZ	−1	4.861	0.255	Estimated by Eq. (7), then fitted by Eq. (102)
Cap pass 2	0	2.589	0.342	
	0.5	1.293	0.562	
S960QL base metal	−1	5.579	9.248	
	0	3.711	0.026	
	0.5	3.208	0.007	

(b) **Material Input Properties**

The base metal and HAZ material data of the materials investigated have already been provided in Sect. 1.9.2 with respect to the cyclic stress-strain curves and the monotonic fracture resistance of S355NL and S960QL and in Sects. 2.2.4.3 and 2.3.2.5.1 with respect to the da/dN–ΔK and fatigue crack propagation threshold data. The coefficients A and b of the cyclic R-curves (Eq. 102 in Sect. 1.3.2.6) have been obtained by curve fits to experimental data or to estimated curves from the modified El Haddad approach (Eq. 7, Sect. 1.4.3). They are summarised in Table 12. For the S355J2+N (3 mm plates) the data of S355NL (10 mm plates) have been used. Although the monotonic stress-strain curves of the base metals of both steels were different, the cyclic stress-strain curves almost coincided (Fig. 112).

(c) **Treatment of Residual Stresses**

A weak point of the analyses so far is that they do not explicitly take into account welding residual stresses although rules for this have been formulated in Sect. 1.5. The reason lies in the uncertainty of the stress state after cyclic loading. This statement must be specified in detail:

(a) There is indication that in the absence of *long range* or reaction residual stresses the as-welded residual stresses at the weld toes are about zero at the surface

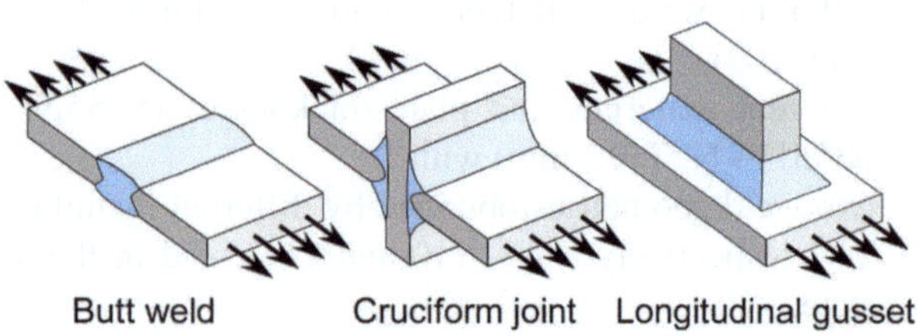

Fig. 109 Weldment types for which finite life (high cycle fatigue) S-N curves have been experimentally generated in IBESS for validation

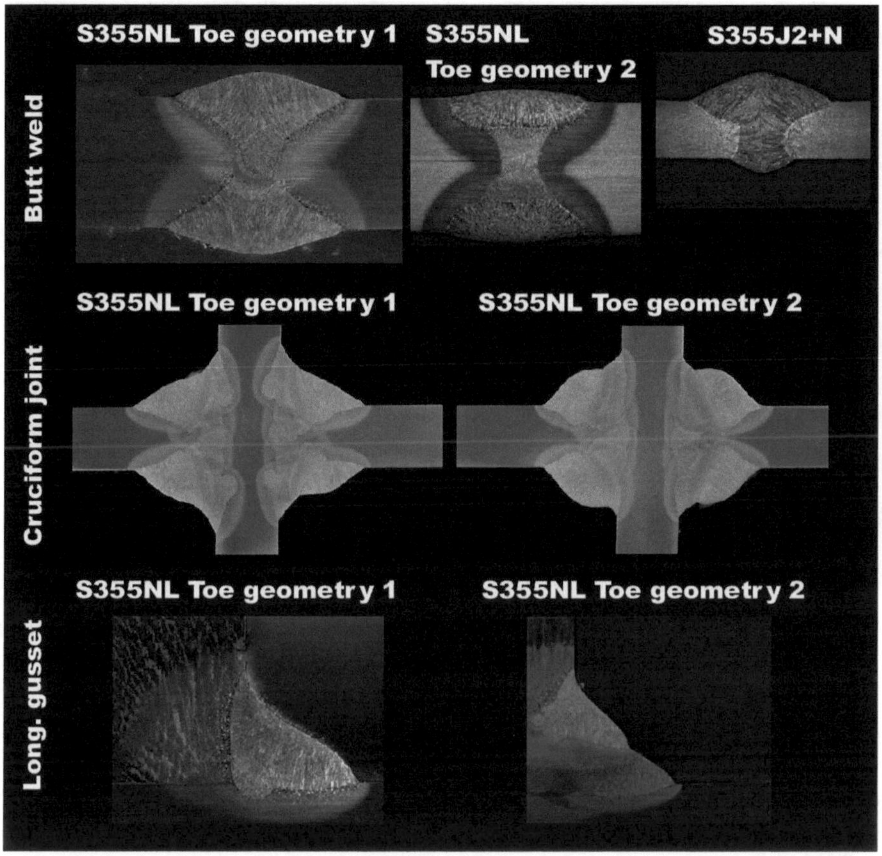

Fig. 110 Macrosections of the weldment configurations and the weld toe geometries investigated within IBESS; steels S355NL and S355J2 + N; according to Schork et al. (2018)

and that even existing tensile or compression residual stresses would relax at cyclic loading at $R = -1$ with cyclic mean stress relaxation also playing a role (see Sect. 1.9.4.2). Thus, an analysis neglecting any welding residual stresses is appropriate at $R = -1$ loading as long as no restraint (long range residual stresses, see Sect. 1.9.4.1) is to be stated. This is the case with respect to the welds investigated here, except the longitudinal gussets.

(b) For higher R ratios ($R \geq 0$), compressive residual stresses have been predicted by finite element analysis for cruciform joints of S355NL (Fig. 36 in Sect. 1.9.4.2, see also Hensel et al. 2018) and for longitudinal gussets of the same steel (Tchuindjang et al. 2018). This makes sense from a mechanical point of view although experimental confirmation is lacking for reasons which are unclear by now (see also Sect. 1.9.4.2). If compressive, i.e., negative residual stresses were built-up at $R = 0$ or higher, the effective R ratio would be lower than the nominal one. In other words: The real mean stress is smaller than those used in the IBESS

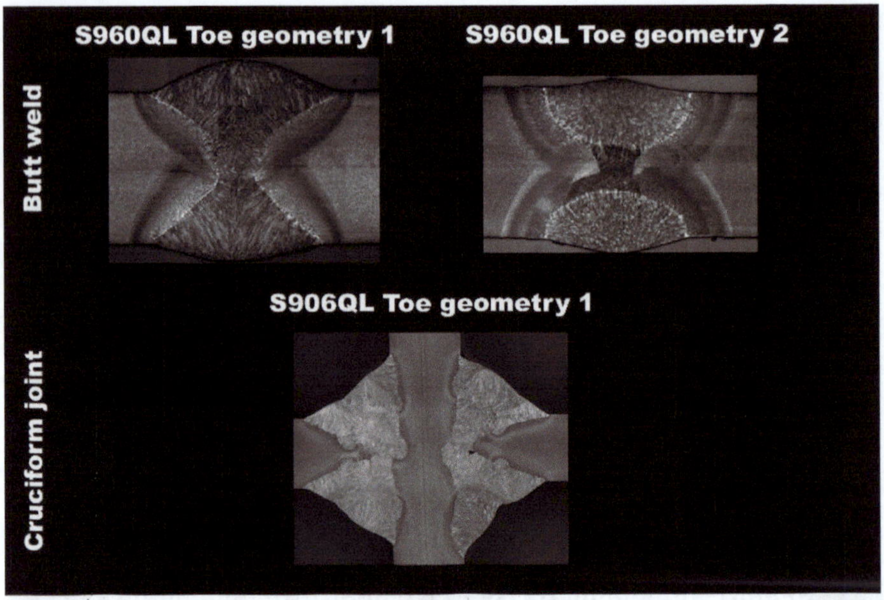

Fig. 111 Macrosections of the weldment configurations and the weld toe geometries investigated within IBESS; steel S960QL; according to Schork et al. (2018)

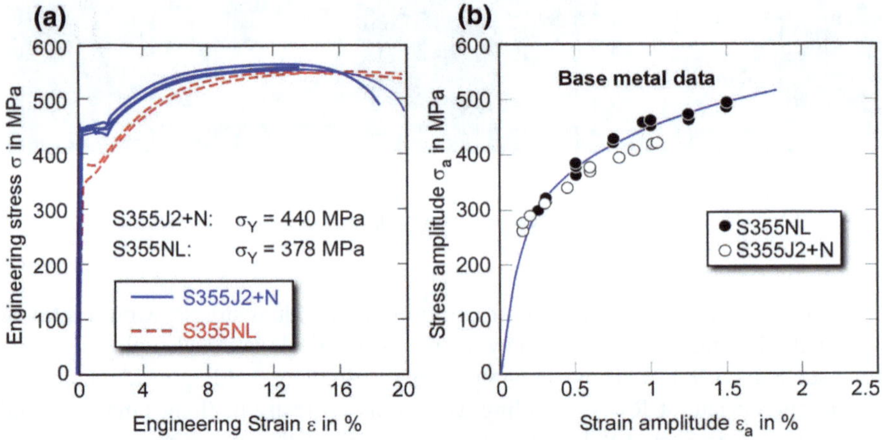

Fig. 112 Monotonic and cyclic stress-strain curves of steel S355J2+N; base metal; according to Kucharczyk et al. (2018)

analysis with the consequence that the latter should be conservative. The effect must be larger for higher than for lower stress ratios because stress redistribution is a question of local plasticity and this will be less pronounced for smaller plastic strains and even not occur under elastic conditions. Note that residual stresses, in an IBESS analysis, are considered by applying a da/dN–ΔK curve for a residual stress corrected, effective R ratio. As a consequence, the IBESS analyses should theoretically be performed for different effective R ratios, R_{eff}, depending on the stress level, i.e., R_{eff} would be lower at higher and higher at lower stress levels.

Note, however, that it was assumed by the preliminary recommendations in Sect. 1.5 that no benefit should be taken from potential compressive stresses which means that all analyses would neglect this kind of residual stresses and, in turn, accept the potential conservatism of the IBESS analysis.

The reason for the preliminary nature of the recommendations of Sect. 1.5 is not only the lack of experimental proof of the compressive stresses when loaded at $R \geq 0$, it is also given by open points in interpreting the numerical evidence. The results in Fig. 36 of Sect. 1.9.4.2 have been obtained for the surface of the weld toe and in the absence of a crack, and the picture might change at and below the surface. The authors also have made use of the stress-strain curve of the base metal instead of those of the HAZ.

Finally, it must be stressed again that all these considerations relate to *medium-range* residual stresses only. *Long-range* or reaction residual stresses have, of course, to be considered in an IBESS analyses. However, with respect to the validation examples of this section, except in the case of the self-restraint longitudinal gussets no reaction residual stresses were present at the as-welded weld toes near the surface and, as mentioned, there is indication of the stress relieving effect of cyclic loading even in this case (Hensel et al. 2018; Tchuindjang et al. 2018).

A total of 31 S-N curves were simulated and compared with experimental data. Selected but representative results are shown in Sect. 2.7.2, for a complete overview see Zerbst (2016). Note that the data are restricted to the finite life (high cycle) fatigue regime of the curves since only these were in the focus of the original IBESS project. Therefore, no validation in the strict sense was performed with respect to the endurance limit. Instead a comparison with a literature compendium of fatigue strength data of structural steels is presented in Sect. 2.7.3. Finally, a first application of IBESS in conjunction with an extended ISO 5187 weld quality class will be provided in Sect. 2.7.4.

2.7.2 Finite Life (High Cycle) Fatigue Regime

2.7.2.1 Butt Welds

In this section, selected representative validation examples are shown. Usually 10 (in some cases 20) simulations have been performed per stress level. The results are shown as scatter bands, i.e., they have not been statistically processed, except those of the few fatigue limits which, however, are not validated here.

A first example has already been presented in Sect. 2.4.6.2, Fig. 103, where butt welds of S355NL loaded at $R = -1$ were investigated. When HAZ data were used,

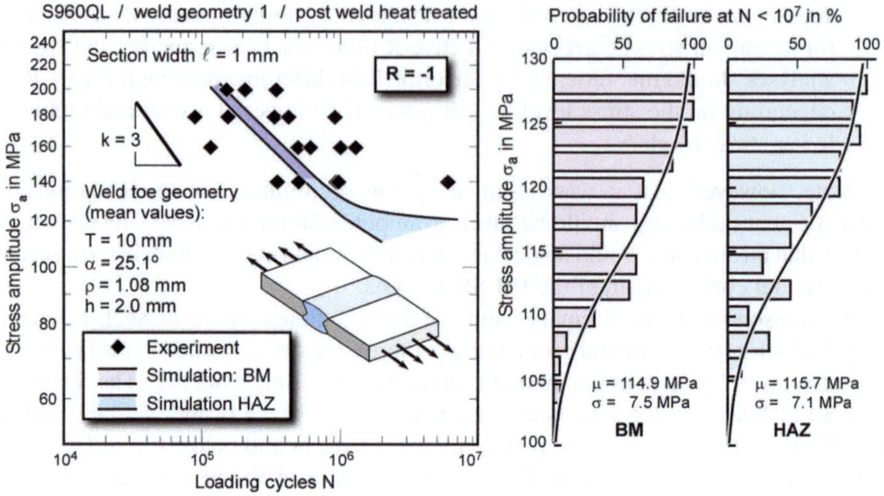

Fig. 113 Finite life (high cycle fatigue) S-N curve and endurance limit prediction; butt weld of steel S960QL, post-weld-heat-treated, $R = -1$. The simulation has been performed for base metal and HAZ properties. The expected values μ and standard deviations σ at the right hand side refer to the fatigue limit defined for $N = 10^7$ loading cycles; according to Madia et al. (2018)

very good predictions have been obtained. Note that this pattern was representative for butt welds loaded at $R = -1$. Another example is shown in Fig. 113, where S960QL butt welds have been investigated. Despite the larger scatter in the experimental results the predictions are again satisfying. Note that so far no scatter of the da/dN–ΔK material data is considered in IBESS.

Focusing on butt welds tested at $R = 0$ or higher the picture changes in that the predictions are conservative, but on a moderate scale. An example is shown in Fig. 114. The authors suspect compressive, i.e., negative residual stresses, are responsible for this, see Sect. 1.9.4.2 and the comments in Sect. 2.7.1. Note that this assumption is confirmed by the investigation of the individual crack in Fig. 115. A similar example has already been shown in Sect. 2.4.3, Fig. 97, where the IBESS analysis for zero residual stresses at the weld toe at $R = -1$ yielded almost perfect results. This time, when performed for nominal $R = 0$, the predictions of the a-N and c-N curves are conservative. For comparison, an $R = -1$ prediction is added which is much closer to the experimental curve at early crack propagation. Later on, this pattern changes and the experimental slopes of the a-N and c-N curves come closer to those of the $R = 0$ predictions. If the assumption of compressive residual stresses at the surface were right, the effective R ratio should be somewhere in between $R = -1$ and $R = 0$. Note that, in contrast to Fig. 97, the calculations started at a crack depth of 2 mm, i.e., only *long* cracks were considered.

Yet another picture emerges in Fig. 116, where predictions for butt welds of 3 mm thick plates of S355J2 + N are compared with experiments at $R = 0$. As noted above, the material properties of S355NL have been assumed for this configuration which

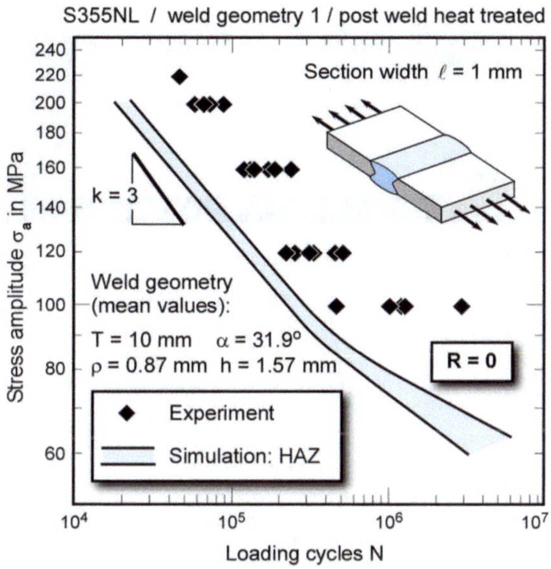

Fig. 114 Finite life (high cycle fatigue) S-N curve; butt weld of steel S355NL, post-weld-heat-treated, R = 0. The simulation has been performed for HAZ properties; according to Madia et al. (2018)

might cause some uncertainty. What is, however, most striking is the small difference between base metal and HAZ based predictions such that both variations seem to be appropriate. Note that the extension of the HAZ is much smaller in the 3 mm-plate compared to the 10 mm one. This is illustrated in Fig. 117 and it allows the conclusion that the *short* crack stage is not confined to the HAZ but a larger part of this takes place within the base metal. With respect to the reduced conservatism compared to Fig. 114 one could suspect that the thin wall prevented or hindered the formation of compressive residual stresses. However, this remains a speculation since no residual stress measurements or simulations were available for this configuration within the project.

An example of an analysis of a butt weld of S960QL loaded at R = 0.5 is shown in Fig. 118. As in the case of S355NL butt welds, the as-welded residual stresses tended to zero at the weld toe, see also Fig. 37 in Sect. 1.9.4.3.1, where the same trend is obtained for a wide range of steels of different strengths, and also in Fig. 38. What is, however, different compared to the lower strength S355NL is that due to the high cyclic yield strength there seems to be little or even no modification of the residual stresses when cyclically loaded. Note that stress redistribution requires plasticity. Therefore, no compressive stresses are to be expected and the analysis with zero residual stresses should give reasonable results. This, despite the large scatter, is reflected in the figure.

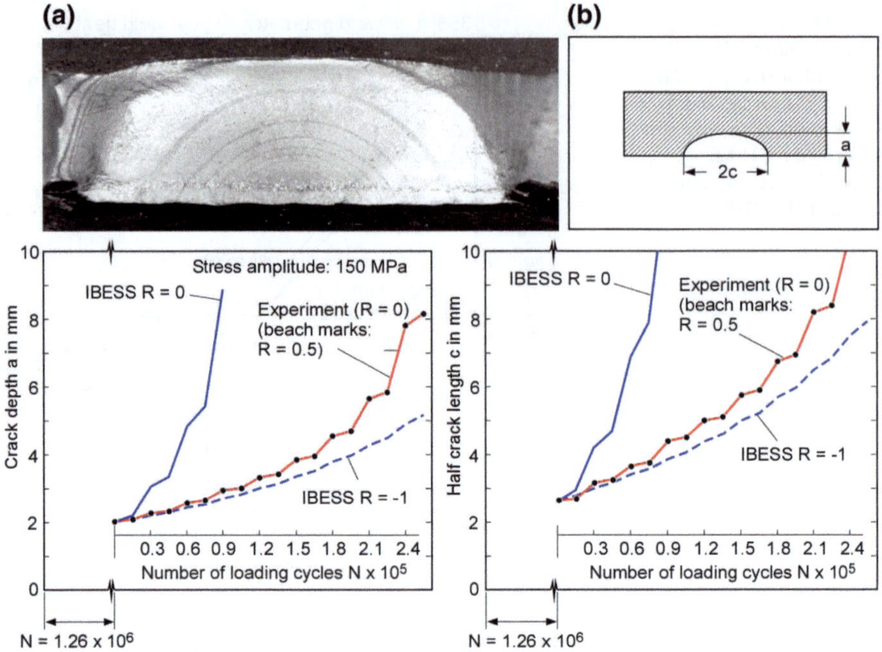

Fig. 115 Propagation of a semi-elliptical surface crack initiated at the weld toe of a cruciform joint at R = 0; steel S355NL. The simulation starts at a crack depth of 2 mm. **a** Development of crack depth, (a); **b** development of (half) crack length at surface, (c). It is suspected that the conservative prediction is due to compressive residual stresses introduced by cyclic loading at R = 0 which are not considered in the calculation; according to Madia et al. (2018)

2.7.2.2 Cruciform Joints

Examples of IBESS analyses on cruciform joints of S355NL and S960QL at R = −1 and 0 are shown in Figs. 119, 120, 121 and 122.

(a) S355NL Cruciform Joints

In contrast to the butt welds of S355NL moderately conservative results are obtained not only for R = 0 (Fig. 120) but also for R = −1 (Fig. 119). The degree of conservatism is almost identical. The reason for this is not clear. One could suspect that different axial misalignments may have played a role with respect to the scatter as well as with respect to the difference between the curves. Note that, as in the case of the butt welds, the as-welded residual stresses at the toe surface of the fillet welds tended to zero. Furthermore, as discussed in Sects. 1.9.4.2 and 2.7.1, finite element predictions yielded compressive residual stresses at the weld toe when loaded at R = 0 such that the effective R ratio is expected to be smaller than the nominal one used for the IBESS prediction.

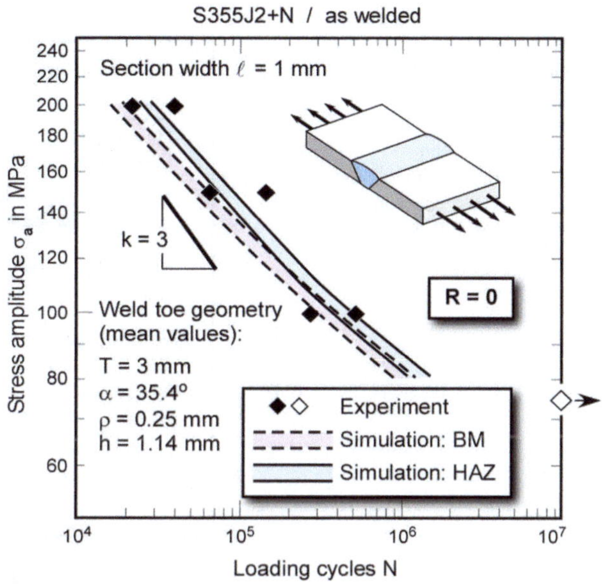

Fig. 116 Finite life (high cycle fatigue) S-N curve prediction; butt weld of steel S355J2+N, Plate thickness: 3 mm; as-welded condition, R = 0. The simulation has been performed for base metal and HAZ properties; according to Madia et al. (2018)

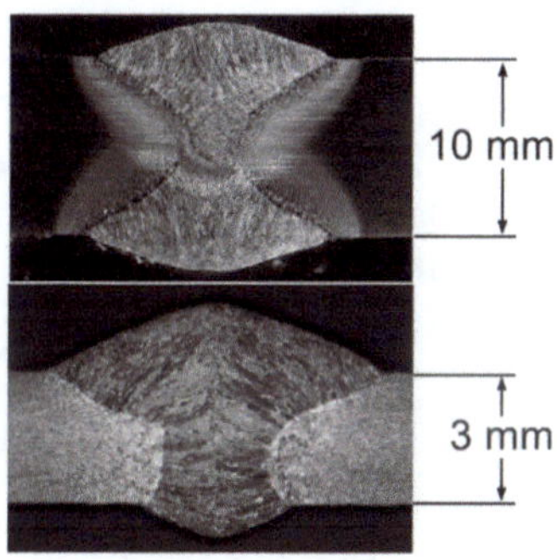

Fig. 117 Macrographs of the S355NL butt welds (plate thickness: 10 mm) and the S355J2+N butt welds (plate thickness: 3 mm). As can be seen, the extension of the HAZ is much smaller for the latter

(b) S960QL Cruciform Joints

The IBESS simulations of these S-N curves are rather close to the experiments, although the R = 0 results are non-conservative at stress levels above $\sigma_a = 160$ MPa (Fig. 122). Note that the corresponding maximum nominal stress in the loading cycle above which the predictions became non-conservative roughly referred to the cyclic yield strengths of the different regions of the HAZ (Fig. 27 in Sect. 1.9.2.2). At that values no analysis would usually be performed.

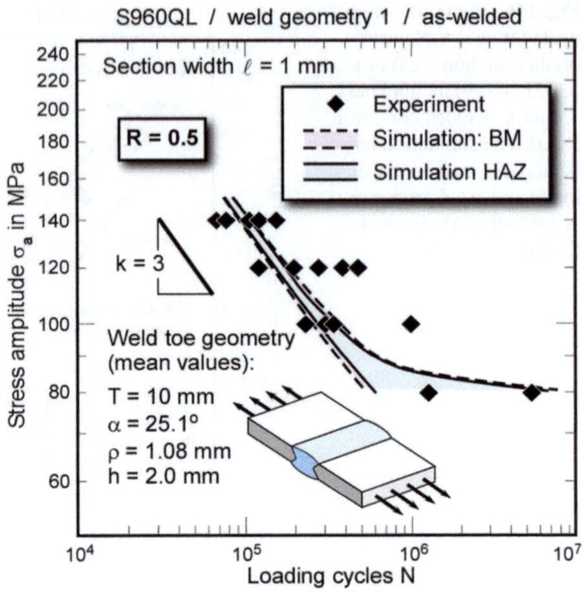

Fig. 118 Finite life (high cycle fatigue) S-N curve prediction; butt weld of steel S960QL, as-welded condition, R = 0.5. The simulation has been performed for base metal and HAZ properties; according to Madia et al. (2018)

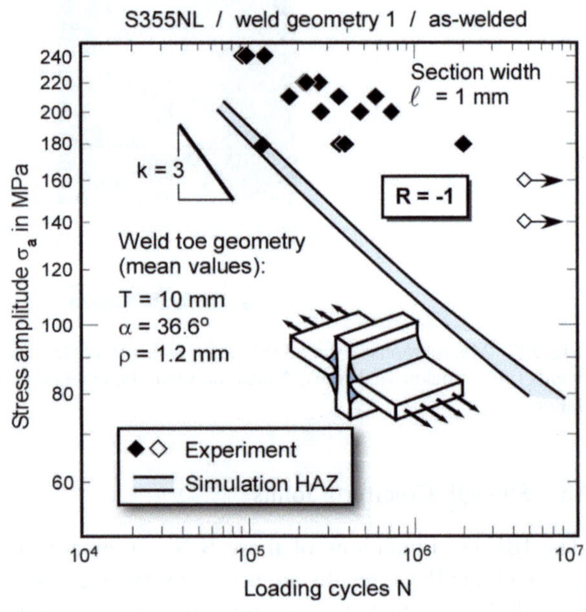

Fig. 119 Finite life (high cycle fatigue) S-N curve prediction; cruciform joint of steel S355NL, as-welded condition, R = −1. The simulation was performed for HAZ properties; according to Madia et al. (2018)

2.7.2.3 Longitudinal Gussets

Analyses of longitudinal gussets are shown in Figs. 123 and 124. For this type of welds no parametric equations had been available for the through-thickness stress

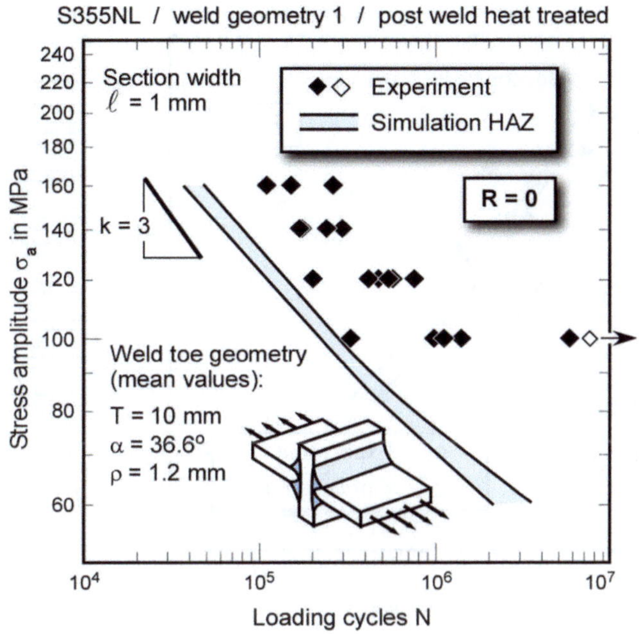

Fig. 120 Finite life (high cycle fatigue) S-N curve prediction; cruciform joint of steel S355NL, post-weld-heat-treated, R = 0. The simulation has been performed for HAZ properties; according to Madia et al. (2018)

Fig. 121 Finite life (high cycle fatigue) S-N curve prediction; cruciform joint of steel S960QL, post-weld-heat-treated, R = −1. The simulation has been performed for HAZ properties; according to Madia et al. (2018)

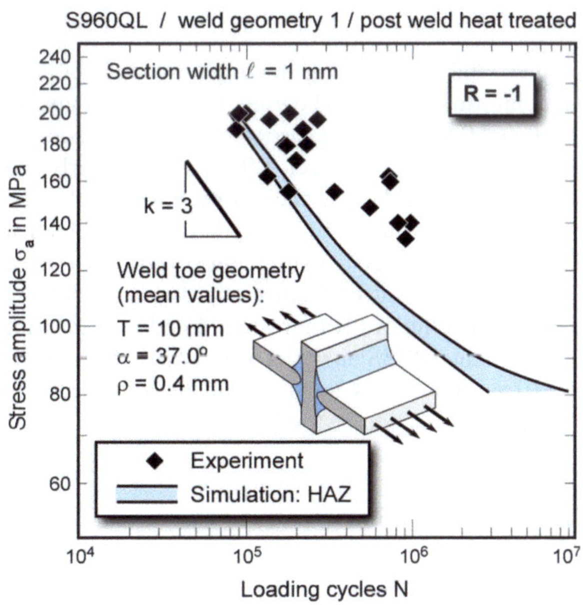

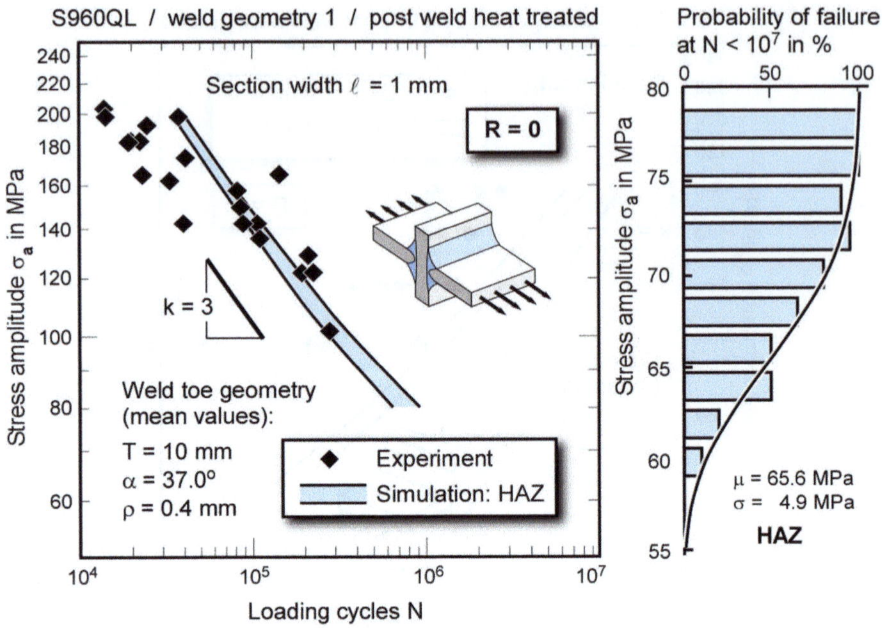

Fig. 122 Finite life (high cycle fatigue) S-N curve and endurance limit prediction; cruciform joint of steel S960QL, post-weld heat treated, R = 0. The simulation has been performed for HAZ properties. The expected value μ and standard deviation σ at the right-hand side refer to the fatigue limit defined for $N = 10^7$ loading cycles; according to Madia et al. (2018)

profiles which vary along the angle to the axial direction. Instead, finite element input information had to be used. Part of this is given in Fig. 125. Both analyses, for R = 0 and 0.5, have been carried out for zero residual stresses which seems to be in conflict with Fig. 34 in Sect. 1.9.4.1 which clearly shows as-welded tensile residual stresses at the weld toe surface. Note, however, that these are expected to relax under cyclic loading with the effect depending on the load amplitude (Hensel et al. 2018). Tchuindjang et al. (2018), simulating the residual stress state in the present specimens by finite elements, even found compressive residual stresses immediately at the weld toe, i.e., the effective R ratio should be lower than the nominal one of R = 0 or 0.5 on which the IBESS analyses were based. As in the case of the butt welds tested at R = 0 (see above), this could explain the conservatism of the analysis: the mean stress is simply assumed to be too high.

2.7.3 Fatigue Limits

As mentioned at the end of Sect. 2.7.1, no validation was performed with respect to the fatigue limit since this was beyond the scope of the original IBESS project. Instead a first comparison will be provided with a comprehensive database on structural

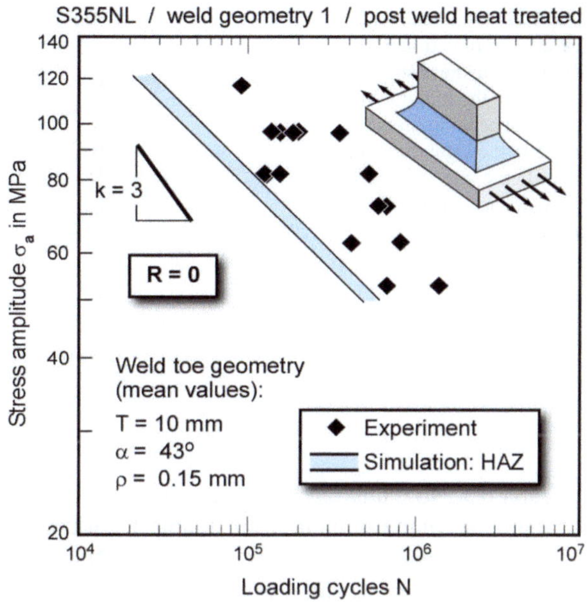

Fig. 123 Finite life (high cycle fatigue) S-N curve prediction; longitudinal gusset of steel S355NL, post-weld heat treated, R = 0. The simulation has been performed for HAZ properties; according to Madia et al. (2018)

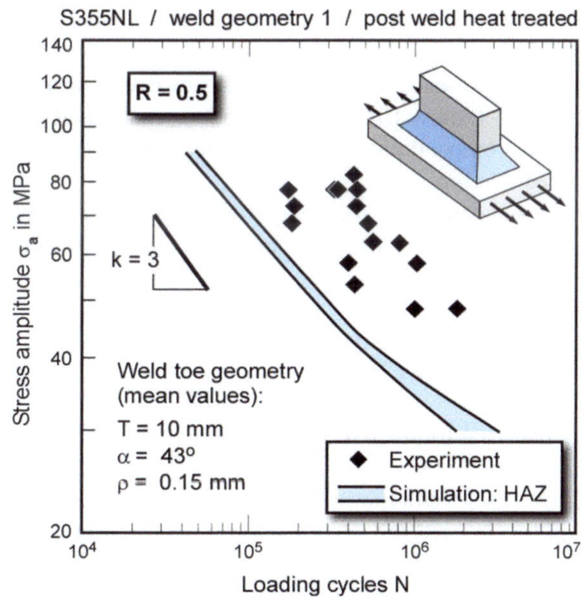

Fig. 124 Finite life (high cycle fatigue) S-N curve prediction; longitudinal gusset of steel S355NL, post-weld heat treated, R = 0.5. The simulation has been performed for HAZ properties; according to Madia et al. (2018)

steels compiled at the University of Darmstadt (Ritter 1994). Based on a very large number of empirical data the author obtained statistically processed stress amplitude ranges at 2×10^6 loading cycles for different weldment types. These are compared with the IBESS lower bound stress amplitudes at the same number of loading cycles

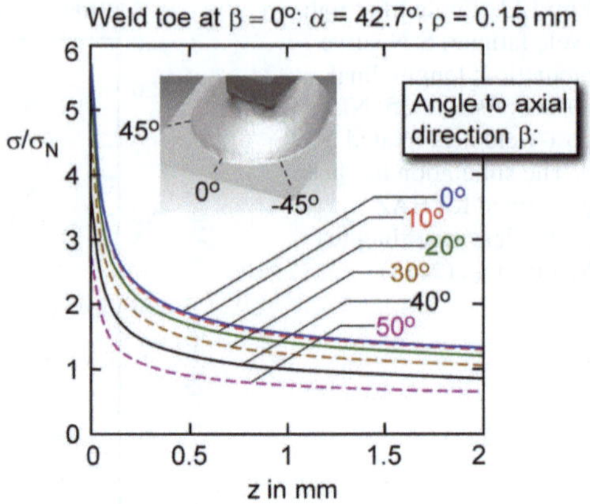

Fig. 125 Finite element based determination of the through-thickness stress profiles in the longitudinal gusset depending on the angle to the axial direction. In this case, the sections ℓ were defined as angular sections

in Table 13. Note that the latter was simply obtained by eye based on lower-bound straight lines of the simulations in double-logarithmic scale. However, as can be seen in the figures in Sect. 2.7.2, the predicted S-N curves deviate from the straight line of the upper finite life (high cycle) fatigue regime in the lower tails which made an easy specification of σ_a at $N = 2 \times 10^6$ difficult in some cases.

As can be seen, the butt weld results of IBESS are within the range of the compendium and this is also the case with respect to the longitudinal gussets with one or perhaps two exceptions which are, however, uncertain. The cruciform joints loaded at $R = -1$ show IBESS lower bound results higher than the 90% percentile value of the compendium whilst the others are in the limits or slightly below the 10% percentile limit.

Note again, that the comparison provided in Table 13 is not a validation in the strict sense since this had to be based on individual experimental results.

2.7.4 IBESS Analysis in Conjunction with an Extended ISO 5817 Weld Quality Class

A first trial to use IBESS in conjunction with an extended weld quality catalogue is provided in Fig. 126. Note that this was not straightforward at any position. E.g., ISO 5817 gives a lower limit for the weld toe radius only for quality group B125, but not for C63 and B90. In these cases, the limit of the parameter was chosen as 1 mm. Different parameters could be chosen with respect to the secondary notch depth, e.g., intermittent undercuts or shrinkage grooves.

A specific problem constituted the very high limit values for the undercut depth given as $0.1 \times T$ but 0.5 mm maximum in ISO 5817 (2014) and $0.1 \times T$ without an additional criterion in Hobbacher (2009). Referring this to a plate thickness of $T = 10$ mm, undercut depths of 0.5 mm and 1 mm were allowed. Instead of these values the average of the measured roughness ($k = P_t = 61.6$ μm) was chosen as

Table 13 Comparison between simulations and statistically specified fatigue limits for structural steels in Ritter (1994)

Geometry	*R*	Compendium (Ritter 1994) σ_a in MPa at N = 2 × 10^6			Simulations σ_a in MPa at N =		Configuration
		10%	50%	90%	2 × 10^6	10^7	
Butt-welded joints	−1	89	109	129	95	79	S355NL WG1 AW
					95	79	S355NL WG1 PWHT
					117	97	S960QL WG1 AW
					117	97	S960QL WG1 PWHT
	0	53	69	85	64	60	S355NL WG1 AW
					64	60	S355NL WG1 PWHT
					70	62	S355NL WG2 AW
					73	68	S355NL WG2 PWHT
					80	75	S355J2+N AW
					–	95	S960QL WG1 AW
					–	95	S960QL WG1 PWHT
					–	119	S960QL WG2 AW
					–	119	S960QL WG2 PWHT
	>0.2	52	64	76	67	60	S355J2+N AW
					–	77	S960QL WG1 AW
Cruciform joints	−1	54	64	74	93	72	S355NL WG1 AW
					93	72	S355NL WG1 PWHT
					83	73	S960QL WG1 AW
					83	73	S960QL WG1 PWHT
	0	52	62	72	58	50	S355NL WG1 AW
					58	50	S355NL WG1 PWHT
					42	35	S355NL WG2 AW
					42	35	S355NL WG2 PWHT
					–	65	S960QL WG1 AW
					–	60	S960QL WG1 PWHT
Longitudinal gussets	0	37	45	53	(34)	30	S355NL WG1 AW
					(40)	30	S355NL WG1 PWHT
					(43)	36	S355NL WG2 AW
					(42)	36	S355NL WG2 PWHT
	>0.2	35	39	43	(28)	25	S355NL WG1 AW
					–	25	S355 WG1 G

WG = weld geometry; AW = as-welded; PWHT = post-weld heat treated. Numbers in brackets = determination uncertain (too few simulations)

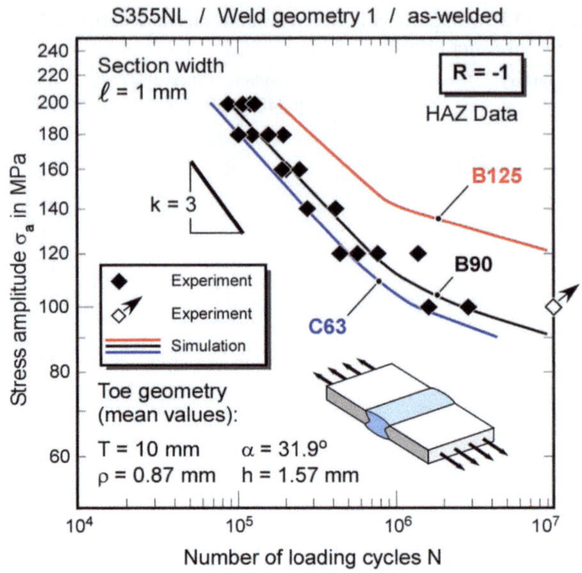

Fig. 126 Application of the IBESS method along with the advanced weld quality criteria according to DIN EN ISO 5817:2014-06, Annex C; steel S355NL butt weld; the experimental S-N data refer to the quality class C63

secondary notch depth for the analysis. The butt welds of S355NL were classified as C63, and it shows up that the IBESS prediction for this was moderately conservative.

The result suggests two conclusions:

(a) At least the IBESS basic procedure is shown to have some potential for application in conjunction with extended quality classes.
(b) This, however, also requires further development of the extended weld quality system in that the permissible undercut depth had to be reduced in the present case.

3 Summary and Future Perspective

The IBESS method presented in the present study is the outcome of a German cluster project finished in 2016. It provides solutions for the fracture mechanics-based determination of the fatigue strength of welds with fatigue crack initiation along the weld toes. To that purpose, it combines established tools for *long* fatigue crack propagation fracture mechanics with novel developments in various fields. These refer to general aspects of fracture mechanics and fatigue strength as well as to specific features of welds. The newly developed elements comprise:

(a) An analytical approach for the cyclic elastic-plastic crack driving force ΔJ which is applicable to the *mechanically short* fatigue crack propagation stage. The method has been validated by finite element analyses.

(b) A method for describing the gradual build-up of the crack closure phenomenon in the *physically short* fatigue crack regime. This is based on the so-called cyclic R-curve, the crack size dependency of the fatigue crack propagation threshold of *physically short* cracks. To that purpose, methodological work has been done in experimental determination of the material data as well as in the development of an estimation method for the cyclic R-curve based on a modified El Haddad approach.

(c) In order to facilitate an integrated analytical analysis, parametric equations have been derived by finite element analyses for stress concentration factors at the weld toe as well as through-wall stress profiles of V and double V butt welds and T and cruciform joints subjected to tension and bending. Compared to existing solutions they are characterized by a wider application field with respect to the toe radius and flank angle and by an improved accuracy when compared to the finite element results.

(d) A method was developed for the determination of the initial crack size needed for fracture mechanics analyses. In contrast with common approaches, where this is based on the adaption to, or on back-calculation from S-N curves, it is based on a two-criteria approach with a lower bound value provided by crack arrest considerations using a cyclic R-curve approach. The initial crack size is provided as a statistical quantity.

(e) Crack arrest analyses are also used for statistically determining a fatigue limit of welded components defined by the stress amplitude referring to 10^7 loading cycles.

(f) At stress levels above the fatigue limit, multiple crack propagation along the weld toe is considered as well as crack coalescence. This is also realized in a stochastic way.

(g) The different growth behavior of the various cracks along the toe is due to a varying local geometry of the latter. This is considered by a semi-stochastic analysis including the statistical distributions of the parameters toe radius, flank angle, weld reinforcement and secondary notch depth with the latter being defined as undercuts, the surface roughness of the base plate near the toe or other items. The statistical distributions were experimentally obtained on a large number of test sets by line scanning and other methods.

(h) Both the fatigue limit as defined above and the finite life (high cycle fatigue) S-N curve are determined as scatter bands and statistical distributions.

(i) An intense discussion is provided on the effect of welding residual stresses. To that purpose measurements and finite element analyses have been performed for the as-welded state as well as for stress redistribution under cyclic loading. As the result, preliminary recommendations have been formulated on how to treat residual stresses in an IBESS analyses.

(j) A large number of finite-life S-N curves have been determined for validation purposes. The data sets comprised different weldment types (butt welds, cross joints and longitudinal gussets), two steels (S355NL and S960QL) of quite different strengths, different weld geometries due to different welding techniques

(TIG, MAG) and as-welded and stress relieved conditions. The results demonstrate the potential of the IBESS method in that it provided good predictions in many cases and moderately conservative ones in others. The reasons for the discrepancies are discussed at the background of potential compressive residual stresses introduced by cyclic loading at R ratios of zero or larger. These have not been considered in the analyses because of a number of uncertain points which need further consideration.

(k) No validation was performed with respect to the fatigue limit because this was outside the scope of the original IBESS project. Instead the results were compared with a large data base on structural steels from the literature, in the vast majority of cases with satisfying results.

(l) In addition to an IBESS basic procedure, a number of simplifications has been proposed some of which, however, lack full validation to date. The aim is a final IBESS procedure consisting in stepwise graded analysis levels.

That brings us to open questions and issue which require further research activity: Besides the remaining validation the most important points are:

(m) Further numerical and experimental investigation on the effect of cyclic loading on the stability of the welding residual stresses particularly when loaded at R ratios ≥ 0.

(n) The combination of IBESS with advanced weld quality criteria. Although the method was shown to provide encouraging results with respect to this, it was also shown that the new weld quality criteria need critical discussion and further development. It is expected that the IBESS method will be capable of providing a valuable contribution to this.

(o) Finally, the method needs some further development. This comprises the introduction of the fatigue crack propagation characteristics in terms of the coefficient C as a further statistical parameter which is not as easy, since reliable statistical data are lacking. Especially the effect of the crack closure phenomenon has to be systematically studied in that context. Another parameter which should be provided as a statistical quantity is the weld width increment L. Finally, the determination and treatment of secondary notch depths is worth further investigation.

Annexes

Introduction to the Annexes

The aim of the Annexes is to provide the user with additional but not exhaustive information. Wherever possible he is referred to documents, procedures and papers dealing with the topics in question.

U. Zerbst et al., *Fatigue and Fracture of Weldments*,
https://doi.org/10.1007/978-3-030-04073-4

Annex A
Master Curve Analyses for the Statistical Determination of Monotonic Fracture Resistance of Welds

Large scatter in monotonic fracture resistance (e.g., in the ductile-to-brittle regime of ferritic materials) is usually described by the VTT-Master curve approach (e.g., Wallin 2002) which is also the topic of ASTM E1921 standard (2010). It is based on a weakest link principle (see Sect. 3.9.5.1) and uses a three-parameter Weibull distribution

$$P = 1 - \exp\left\{-\left[\frac{(K_{mat} - K_{min})}{K_0 - K_{min}}\right]^m\right\} \tag{A.1}$$

with P being the failure probability (of the test specimens) and K_{mat} the fracture toughness in terms of the K-factor, which is formally determined from the critical J-integral, J_{mat} by

$$K_{mat} = \sqrt{J_{mat} \cdot E / (1 - \nu^2)} \tag{A.2}$$

K_0, K_{min} and m are the scale, the shift and the shape parameter of the distribution. In the general case, these are all fit parameters, but in the Master Curve concept, the application range of which is restricted to ferritic steels with yield strengths between $\sigma_Y = 275$ and 825 MPa, the shape parameter is fixed as m = 4 and the shift parameter as $K_{min} = 20\ \text{MPa}\sqrt{\text{m}}$, both on empirical basis.

Note that the concept has been formally applied to the HAZs of the investigated S355NL and S960QL steel welds in Sect. 3.9.2.3 (which both showed a martensitic-bainitic microstructure) and, as can be seen from Figs. 29b–30b the fits were not perfect. A possible option in such cases, which is however not covered by a test standard, is to fit the data with another shift factor K_{min} than $20\ \text{MPa}\sqrt{\text{m}}$. An example is provided by Romano et al. (2016) for rail steel.

With a transition temperature T_0 referring to a median (50% probability) fracture toughness of $100\ \text{MPa}\sqrt{\text{m}}$ determined with 1T specimens (thickness ≈ crack front

U. Zerbst et al., *Fatigue and Fracture of Weldments*,
https://doi.org/10.1007/978-3-030-04073-4

length = 1 inch ≈ 25 mm), a Master Curve based statistical transition curve for ferritic steels is obtained:

$$K_{med} = 30 + 70\ \exp\left[0.019\ (\mathrm{T} - \mathrm{T_0})\right]\ \text{in MPa} \cdot \mathrm{m}^{1/2}$$
$$(\text{validity range: } -50\,^\circ\mathrm{C} \leq \mathrm{T} - \mathrm{T_0} \leq +50\,^\circ\mathrm{C}). \tag{A.3}$$

Equations (A.1) and (A.3) have been developed for (macroscopic) homogeneous material, a condition which is obviously not fulfilled for weldments, see Sect. 3.9.2. The impact of the material inhomogeneity on statistical treatment is schematically illustrated in Fig. A.1, where the weld is treated in a simplified manner as a two-material system. Both "sub-materials" are characterized by their own scatter bands and statistical distributions. When the data points of both materials are mixed and treated as one sample, obviously none of the original distributions is correctly reproduced. What is of particular importance, is that the fitted curve (dashed line) overestimates the fracture resistance at the lower tail of the distribution which refers to the less tough material, i.e., usually the HAZ in a weld which is most relevant for assessment.

What, therefore, is needed, is a procedure for "demixing" the various data sub-sets. To this end, Wallin et al. (2004) developed two methods, a lower tail Master curve procedure and a bimodal master curve method (which unlike the former is also able to describe the toughness at the upper tail of the distribution). The first one became part of the SINTAP procedure (1999), see also Zerbst et al. (2007) and was taken over in Annex L of BS7910 (2013). It will not be outlined

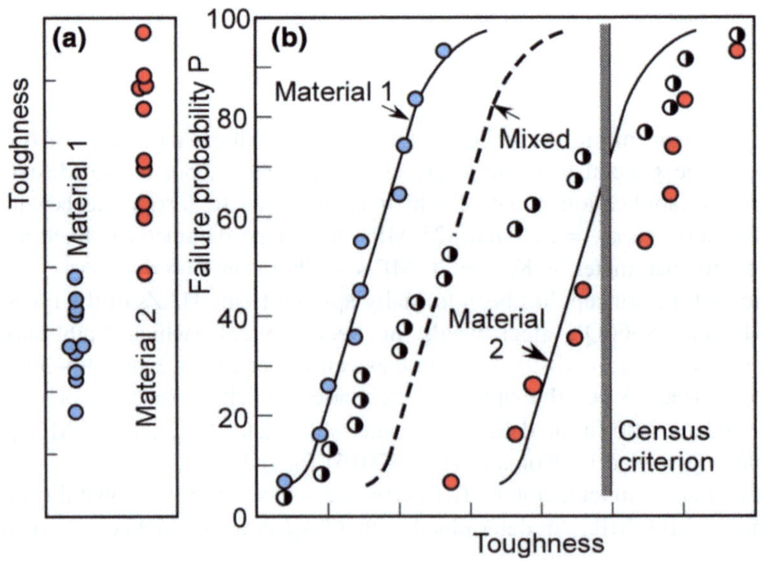

Fig. A.1 Schematic illustration of the effect of inhomogeneous material such as across a weld on the statistical distribution of the monotonic fracture resistance; according to Zerbst et al. (2007)

here in detail, see, however, the descriptions in the papers cited and in Appendix A of Zerbst et al. (2014a) where it has been reproduced from Zerbst et al. (2007). The basic principle is that a censoring criterion (i.e., the maximum toughness value to be included in the analysis) is step-wise decreased. Comparing the results of the different steps, information is gathered on the impact of the material inhomogeneity on the lower tail distribution. When the latter becomes stable, i.e. the toughness for a certain probability no longer decreases when the censoring level is lowered, the analysis is stopped and the resulting lower tail of the distribution can be used for the assessment.

Annex B
Strength Mis-match η_p Factors for J-Integral Determination

A strength mis-match corrected J-integral based on a mis-match η_p factor has been introduced in Sect. 3.9.5.2. Solutions for this are, e.g., provided by Schwalbe et al. (2002), Appendix 9. These are reproduced in Fig. B.1 for different specimen geometries, notch positions and strength mis-match ratios. Further information is found in Kim and Budden (2001), Kim (2002), Kim et al. (2003), Paredes and Ruggieri (2012) and Koo et al. (2012), see also the review paper of Zerbst et al. (2014a).

U. Zerbst et al., *Fatigue and Fracture of Weldments*,
https://doi.org/10.1007/978-3-030-04073-4

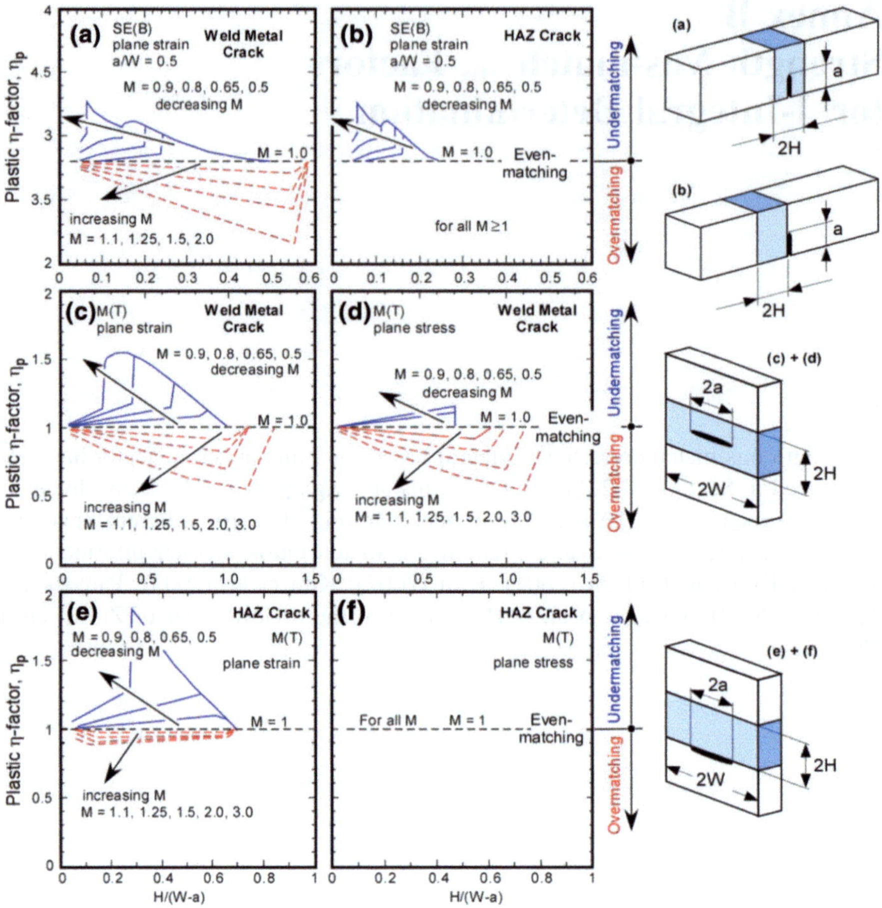

Fig. B.1 Plastic η factors, η_p, of welded specimens with strength mis-match, according to Schwalbe et al. (2002); results from Kim et al. (2003)

Annex C
Determination of the Monotonic Elastic-Plastic Crack Driving Force Including Strength Mis-match and Secondary Stresses

C.1 Basic Principle

The monotonic elastic-plastic crack driving force can be determined by finite elements or—more suitable in conjunction with IBESS—by analytical methods such as R6 (2014), BS7910 (2013) or SINTAP (1999), see also Zerbst et al. (2007) and Zerbst and Madia (2018a). The basic principle is that the crack driving force is determined as a stress intensity factor (see Sect. 4.2.3.2.5, however for monotonic loading, i.e. K is determined instead of ΔK) which is then corrected for ligament plasticity by a function $f(L_r)$. The monotonic counterparts to the equations given in Sect. 4.2.2.2 are

$$J = \frac{K^2}{E'} \cdot [f(L_r)]^{-2}, \tag{C.1}$$

$$L_r = \frac{\sigma_{app}}{\sigma_0} \text{ and} \tag{C.2}$$

$$f(L_r) = \left[\frac{E \cdot \varepsilon_{ref}}{\sigma_{ref}} + \frac{1}{2}\frac{L_r^2}{E \cdot \varepsilon_{ref}/\sigma_{ref}}\right]^{-1/2}. \tag{C.3}$$

For the determination of the reference yield stress σ_0 see Sect. 2.2.3.2.6. Instead of using Eq. (C.2), $f(L_r)$ can also be determined by

$$L_r = \frac{\sigma_{ref}}{\sigma_Y} \tag{C.4}$$

as in BS7910 or based on a yield or limit load F_Y

U. Zerbst et al., *Fatigue and Fracture of Weldments*,
https://doi.org/10.1007/978-3-030-04073-4

$$L_r = \frac{F}{F_Y} \tag{C.5}$$

as, e.g., in the ETM procedure of the former GKSS Research Centre (Schwalbe et al. 1998). Solutions are found in various compendia such as Miller (1988), Schwalbe et al. (1997, 1998), Laham (1999), Dillström et al. (2008), R6 (2014), BS7910 (2013) and others.

Note that Eq. (C.3) is the $f(L_r)$ function of the highest analytical option in R6, BS7910 and SINTAP but lower options can be applied as well. Equation (C.1) follows what in SINTAP is designated as a CDF approach (the acronym CDF stands for "crack driving force"). The crack driving force (usually in terms of J) in the component is determined as a function of applied load and subsequently compared with the monotonic fracture resistance of the material.

In contrast, the FAD approach (the acronym FAD stands for "failure assessment diagram") uses the crack driving force (in terms of the K-factor) as referred to the fracture resistance ($K_r = K/K_{mat}$). Its basic equation is

$$K_r = \frac{K}{K_{mat}} \cdot f(L_r). \tag{C.6}$$

Again, $f(L_r)$ provides a correction of the crack driving force for ligament yielding. Figure C.1 illustrates the use of the failure assessment diagram. The function $f(L_r)$ defines a failure line, the FAD, which is assumed to be a function of the material but independent of the component geometry and loading type (which is a moderate conservative simplification). The assessment is based on the relative position of a design point (L_r, K_r) for the component to be evaluated. Its position depends on the component geometry, the loading type, the crack size and geometry and the yield strength of the material. As long as the assessment point is inside the area circumscribed by the FAD, the component is regarded as safe, when it crosses this line it becomes potentially unsafe. Load increase or a larger crack size will usually move the assessment point to higher L_r as well as higher K_r and, thus, towards the FAD line (Fig. C.1a, b). In case of unstable crack extension following a phase of stable crack growth things are less intuitively accessible because in addition to the K-factor and L_r also the fracture resistance K_{mat} (as the second term in K_r) becomes a function of the crack size. As the consequence, the path of the assessment point follows the pattern shown in Fig. C.1c. The transition from stable to unstable crack extension is predicted for those load the curve of which touches the FAD curve tangentially.

If a CDF approach is used, instability is predicted by a monotonic R-curve analysis which already has been briefly explained in Sect. 4.4.1 because of its similarity with the cyclic R-curve analysis introduced there. The monotonic counterpart is illustrated in Fig. C.2. At the load the crack driving force curve (J-a)

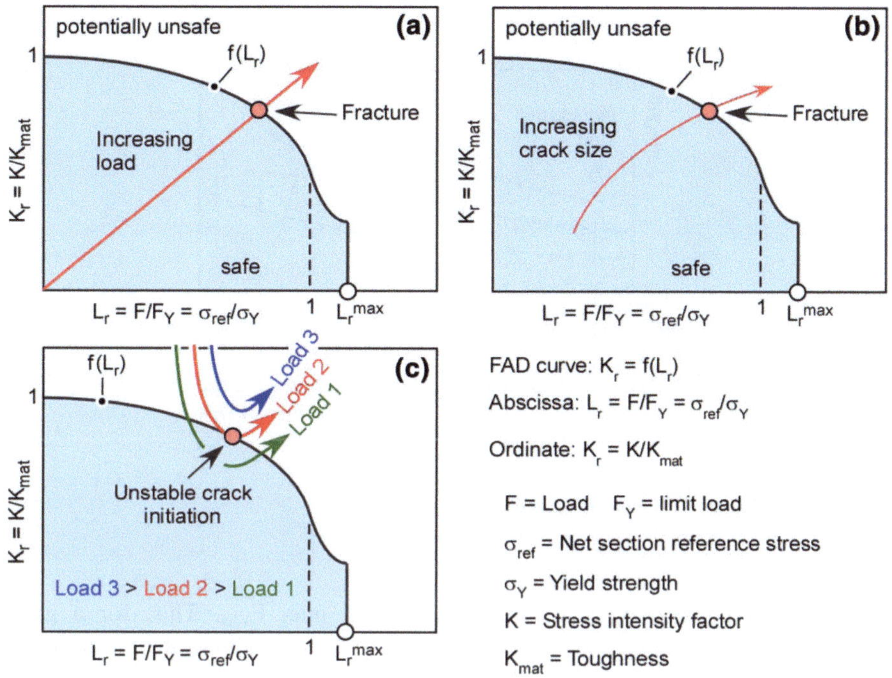

Fig. C.1 FAD analyses. **a** For increasing applied load; **b** For increasing crack size; **c** Instability analysis

of which fulfills the tangency criterion with the monotonic R-curve (J-Δa) for a certain pre-existing crack depth (a_0), initial stable crack extension will turn to unstable crack growth.

C.2 Treatment of Strength Mis-match

Strength mis-match has been the topic of Sect. 3.9.3, its treatment in material fracture resistance that of Annex B. Here, an analytical approach for taking into account the effect in monotonic fracture analysis as it is used in Schwalbe et al. (1997), SINTAP (1999) and Zerbst (2007), Appendix IV.2 of R6 (2014) and Annex I of BS7910 (2013) shall be briefly introduced. Only the basic lines will be explained, for detailed information the reader is referred to the sources cited, see also Zerbst et al. (2014a).

Compared to the homogenous condition, strength mis-match, i.e. adjacent material sections show different strengths, influences the patter of yielding, see Fig. 32 in Sect. 3.9.3. This is taken into account by a mis-match corrected ligament yielding parameter L_r. E.g., using the definition of L_r in Eq. (C.5) the yield or limit

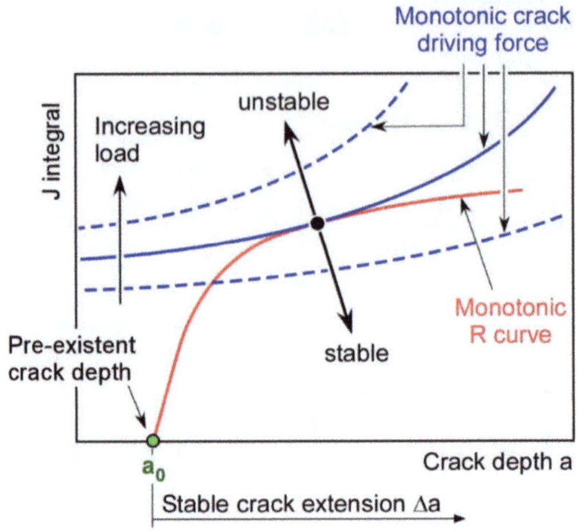

Fig. C.2 CDF analysis for crack instability

load F_Y is preplaced by a mis-match corrected one, F_{YM}. This, for a given geometry, depends on the mis-match ratio M (Eq. 9), the width of the material strip with the different strength (H or 2H), the location of the crack with respect to the interface between the different areas and the crack size a. An example is provided in Fig. C.3, where F_{YB} is the yield or limit load of the base material.

A problem on its own is the definition of the strip width H for real welds which, e.g., show a V, double-V or other shape. Preliminary proposals are given in Fig. C.4.

The common definition of the mis-match ratio M has been provided by Eq. (9) in Sect. 3.9.3. In a weldment, it refers to the weld and base metal. However, reality is usually more complex. Frequently the strength of the HAZ as a third region is even higher than that of the base and weld metals in steels and sometimes there is even more detailed graduation. However, only two-material solutions for F_{YM} or similar parameters are available in the documents summarized above. Nevertheless, these solutions can often be used by simplifying multi-material to two-material configurations when M is defined in a more flexible way for the two material sections under consideration.

A schematic overview of when and how strength mis-match is expected to play a role for the crack driving force such that mis-match corrected L_r factors should be used is provided in Fig. C.5. The phrase "geometry and loading mode dominant" refers to potential constraint effects which will not be addressed here, see however, Revision 4 of R6 (2014) and Annex N of BS7910 (2013) and also Zerbst and Madia

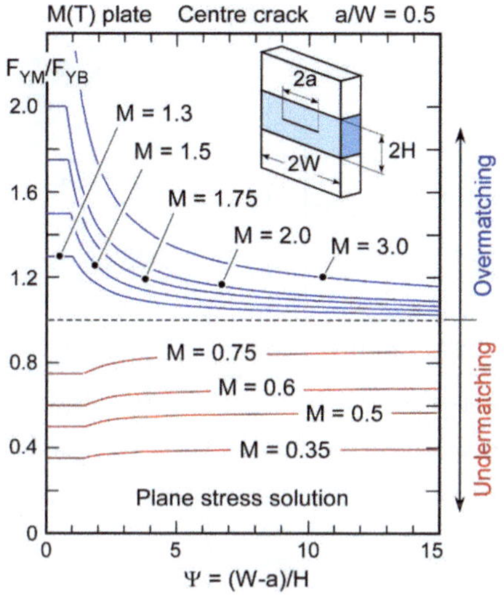

Fig. C.3 Strength mis-match corrected yield or limit load F_{YM} as referred to its base plate equivalent F_{YB}; according to Schwalbe et al. (2007)

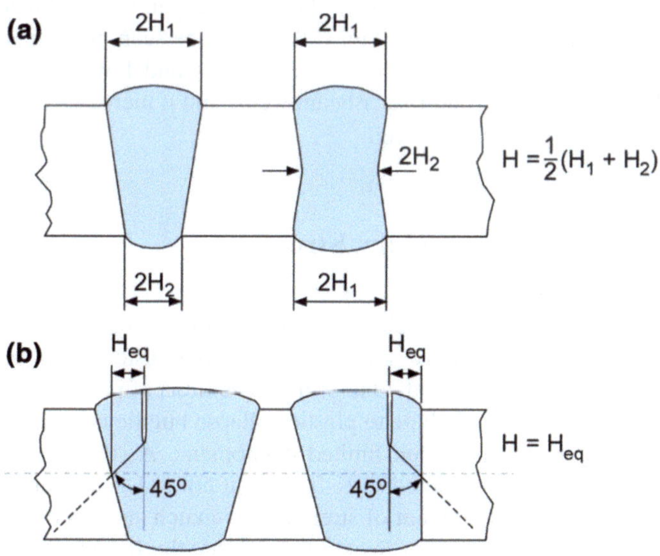

Fig. C4 Definition of the weld strip width 2H or H in cases where the weld does not have a simple prismatic shape. **a** proposal according to ETRM GTP (Schwalbe et al. 2002); **b** Proposal according to Junghans (1998). The equivalent value H_{eq} is defined on the basis of the shortest distance between the crack tip and the fusion line along the slip lines emanating from the crack tip

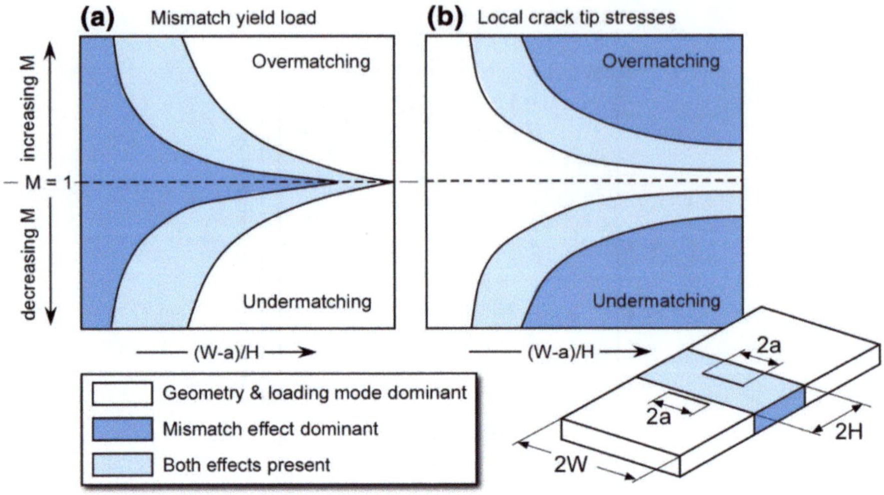

Fig. C.5 Schematic diagram showing the dominance of strength mis-match and geometry/loading on **a** the yield or limit load, **b** local crack tip stresses/stress triaxiality; according to Schwalbe et al. (1997)

(2018a). The effect of strength mis-match on the yield or limit load as well as on the local crack tip stresses is illustrated in Fig. C.3 and in Fig. 44 in Sect. 3.9.5.2. Note that the effect on stress triaxiality shows the reversed pattern combined to that on the mis-match yield or limit load in Fig. C.5. As can be seen, the effect of strength mis-match decreases for high W-a, i.e. small cracks and low H, i.e. small weld widths such as, e.g. in electron or laser beam welds and it increases with both, more pronounced under and overmatching.

C.3 Treatment of Secondary Stresses

The topic of secondary stresses including its definition has been discussed in some detail in Sect. 3.9.4. Here, only some notes on its consideration in monotonic fracture assessment shall be added. One of the definitions of primary and secondary stresses is that the former contribute to plastic collapse but the latter don't since they arise from displacement and strain limited phenomena. As the consequence, secondary stresses are only considered in K-factor but not in L_r determination. This is exactly the reverse of the treatment of strength mis-match such as discussed above.

It has already been discussed in Sect. 3.9.4.3.3 that the K-factors due to primary and secondary stresses, K_I^p and K_I^s, cannot be simply superimposed because of plasticity and relaxation effects. Instead a plasticity correction factor V to K_I^s has to be applied which depends on the degree of ligament yielding and other parameters (Fig. 40). With this, Eq. (C.1) has to be modified to

$$J = \frac{\left(K_I^p + V \cdot K_I^2\right)^2}{E'} \cdot [f(L_r)]^{-2} \tag{C.7}$$

for the CDF approach and Eq. (C.6) has to be modified to

$$K_r = \frac{K_I^p + V \cdot K_I^s}{K_{mat}} \cdot f(L_r) \tag{C.8}$$

for the FAD approach. With respect to the determination of V the authors refer to Revision 4 of R6 as the most up-to-date document in the field, for some discussion see also the notes in Sect. 3.9.4.3.3 and in Zerbst et al. (2014a).

Annex D
Roughness Determination Close to the Weld Toe for Specifying the Secondary Notch Depth k

Roughness measurement in IBESS is based on ISO 4287 (1997), however, with some modification. From the large number of roughness parameters defined there, the measure P_t is chosen for specifying the secondary notch depth k near the weld toe. This characterizes the maximum height of the assessed profile. The main difference to the standard is that only a close range next to the weld toe is considered such as illustrated in Fig. D.1. Note that this parameter varies along the weld toe length. In common roughness determination, a larger measuring length is taken as the basis and the raw data are smoothed, i.e., large-scale waviness of the surface is removed by electronic filtering or by subsequent processing of the data in the computer. Since only the weld toe is of interest in the context of IBESS, unfiltered and unsmoothed data of the region are used.

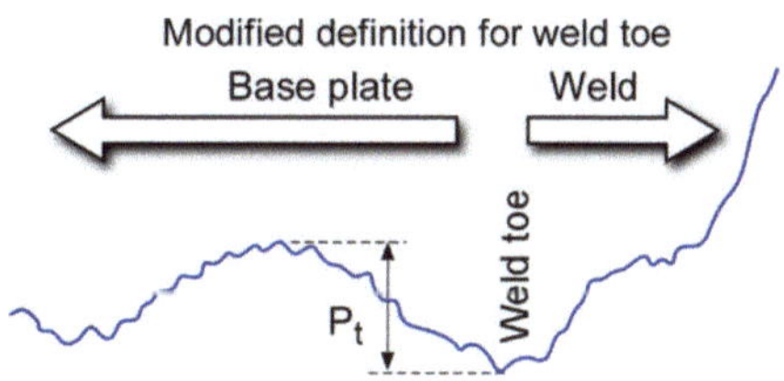

Fig. D.1 Determination of the roughness parameter P_t of the primary profile at the weld toe

U. Zerbst et al., *Fatigue and Fracture of Weldments*,
https://doi.org/10.1007/978-3-030-04073-4

Literature

Ainsworth, R. A. (1984). The assessment of defects in structures of strain hardening material. *Engineering Fracture Mechanics, 19,* 633–642.

Akinawa, Y., Zhang, L. M., & Tanaka, K. (1997). Prediction of the fatigue limit of cracked specimens based on the cyclic R-curve method. *Fatigue & Fracture of Engineering Materials & Structures, 20,* 1387–1398.

Anderson, T. L. (2003). Flaw characterization. In R. A. Ainsworth, K.-H. Schwalbe, (Eds.), *Comprehensive structural integrity* (Vol. 7, pp. 227–243). Practical Failure Assessment Methods, Elsevier.

ASTM E 140-12. (2012). Standard hardness conversation tables for metals relationship among Brinell hardness, Vickers hardness, Rockwell hardness, Superficial hardness, Knoop hardness, Scleroscope hardness, and Leeb hardness., West Conshohocken, PA: American Society for Testing and Materials.

ASTM E 647-11. (2011). *Standard test method for measurement of fatigue crack growth rates.* Philadelphia: American Society for Testing and Materials (ASTM).

ASTM E 1921-10. (2010). Standard test method for determination of reference temperature, T_0, for ferritic steels in the transition range. West Conshohocken, PA: American Society for Testing and Materials.

Atzori, B., Meneghetti, G., & Susmel, L. (2005). Material fatigue properties for assessing mechanical components weakened by notches and defects. *Fatigue & Fracture of Engineering Materials & Structures, 28,* 83–97.

Barsoum, Z., & Jonsson, B. (2011). Influence of weld quality on the fatigue strength in seam welds. *Engineering Failure Analysis, 18,* 971–979.

Basan, R., Franulovic, M., & Smokivina Hanza, S. (2010). Estimation of cyclic stress-strain curves for low alloy steel from hardness. *Metallurgia, 49,* 83–86.

Bate, S. K., Green, D., & Buttle, D. (1997). A review of residual stress distributions in welded joints for the defect assessment of offshore structures. Health and Safety Executive (HSE), Offshore Technology Report.

Beghini, M., Bertini, L., & Vitale, E. (1994). Fatigue crack growth in residual stress fields: Experimental results and modelling. *Fatigue & Fracture of Engineering Materials & Structures, 17,* 1433–1444.

Beretta, S., Carboni, M., & Madia, M. (2006). Fatigue strength in presence of inhomogeneities: Influence of constraint. *Journal of ASTM International, 3*(4).

Beretta, S., Carboni, M., & Madia, M. (2009). Modelling of fatigue thresholds for small cracks in a mild steel by "Strip-Yield" model. *Engineering Fracture Mechanics, 76,* 1548–1561.

U. Zerbst et al., *Fatigue and Fracture of Weldments*,
https://doi.org/10.1007/978-3-030-04073-4

Björk, T., Samuelsson, J., & Marquis, G. (2008). The need for a weld quality system for fatigue loaded structure. *Welding in the World, 52,* 34–46.

Brennan, F. P., Peleties, P., & Hellier, A. K. (2000). Predicting weld toe stress concentration factors for T and skewed T-joint plate connections. *International Journal of Fatigue, 22,* 573–584.

Breßler, S. (2014). Ermittlung von Spannungstiefenprofilen von auf Zug-und Biegung belasteten Stumpfschweißnähten in Stahl mittels FEM und Parametergleichungen (Master Thesis). Berlin (in German).

Brook, R. (1983). Cited in Taylor, D. (1989): *Fatigue Thresholds.* Butterworth & Co Publishers.

Bruzzi, M. S., & McHugh, P. E. (2002). Methodology for modelling the small crack fatigue behaviour of aluminium alloys. *International Journal of Fatigue, 24,* 1071–1078.

BS 7448. (1997). *Fracture mechanics toughness tests. Part 2: Method for determination of KIc, critical CTOD and critical J values of welds in metallic materials.* London: The British Standards Institution (BSI) Standards Publication.

BS 7910. (2013). *Guide to methods for assessing the acceptability of flaws in metallic structures.* Including Amendment (2015) and Corrigenda I-2. London: The British Standards Institution (BSI) Standards Publication.

Budden, P. J., & Sharples, J. K. (2003). Treatment of secondary stresses. In R. A. Ainsworth, K.-H. Schwalbe (Eds.), *Comprehensive structural integrity* (Vol. 7, pp. 245–287). Practical Failure Assessment Methods, Elsevier.

Çam, G., Koçak, M., & Dos Santos, J. (1999). Developments in laser welding of metallic materials and characterization of the joints. *Welding World, 43,* 13–25.

Capelli, M. D., Carlson, R. L., & Kardomates, G. A. (2008). The transition between small and long fatigue crack behavior and its relation to microstructure. *International Journal of Fatigue, 30,* 1473–1478.

Carboni, M., & Regazzi, D. (2011). Effect of the experimental technique onto R dependence of ΔKth. *Procedia Engineering, 10,* 2937–2942.

Chapetti, M. D., Belmonte, J., Tagawa, T., & Miyata, T. (2004). Integrated fracture mechanics approach to analyse fatigue behavior of welded joints. *Science and Technology of Welding and Joining, 9,* 430–438.

Chen, J. H., & Cao, R. (2015). *Micromechanisms of cleavage of metals. A comprehensive microphysical model for cleavage cracking in metals.* Amsterdam: Elsevier.

Davidson, D., Chan, K., McClung, R., & Hudak, S. (2003). Small fatigue cracks. In R. O. Ritchie, Y. Murakami (Eds.), *Comprehensive structural integrity* (Vol 4, pp. 129–164). Cyclic loading and Fracture, Elsevier.

Dawes, M. G., Pisarski, H. G., & Squirrell, S. J. (1989). *Fracture mechanics tests on welded joints.* ASTM STP 995 (pp. 191–213). Philadelphia: American Society for Testing and Materials (ASTM).

Dillström, P., Bergman, M., Brickstad, B., Zang, W., Sattari-Far, I., Andersson, P., et al. (2008). *A combined deterministic and probabilistic procedure for safety assessment of components with cracks—handbook.* Swedish Radiation Safety Authority.

Dillström, P., & Nilsson, F. (2003). Probabilistic fracture mechanics. In R. A. Ainsworth, K.-H. Schwalbe (Eds.), *Comprehensive structural integrity* (Vol 7, pp. 467–489). Practical Failure Assessment Methods, Elsevier.

Dos Santos, J., Çam, G., Torster, F., Isfan, A., Riekehr, S., Ventzke, V., et al. (2000). Properties of power beam welded steels, Al- and Ti-alloys: Significance of strength mismatch. *Welding World, 44,* 42–64.

Dowling, N. E. (1979). Notched member fatigue life predictions combining crack initiation and propagation. *Fatigue & Fracture of Engineering Materials & Structures, 2,* 129–138.

Dowling, N.E. (1998). *Mechanical behaviour of materials.* (2nd ed). ed., App. B: Statistical Variation in Material Properties, Prentice Hall Inc.

Dowling, N. E., & Begley, J. A. (1976). *Fatigue crack growth during gross plasticity and the J integral*. ASTM STP 590 (pp. 82–103). Philadelphia: American Society for Testing and Materials (ASTM).

El Haddad, M. H., Smith, K. N., & Topper, T. H. (1979). Fatigue crack propagation of short cracks. Trans. ASME. *Journal of Engineering Materials and Technology, 101,* 42–46.

Endo, M., & McEvily, A. J. (2007). Prediction of the behavior of small fatigue cracks. *Materials Science and Engineering: A, 468–470,* 51–58.

Farajian, M. (2013). Welding residual stress behavior under mechanical loading. *Welding World, 57,* 157–169.

Fett, T., Munz, D., & Neumann, J. (1890). Local stress intensity factors for surface cracks in plate under power-shaped stress distributions. *Engineering Fracture Mechanics, 36,* 647–651.

Forman, R. G. & Mettu, R. R. (1992). *Behaviour of surface and corner cracks subjected to tensile and bending loads in Ti-6Al-4V alloy*. ASTM STP 1131 (pp. 519–546). Philadelphia: American Society for Testing and Materials (ASTM).

Gagg, C. R. (2005). Failure of components and products by 'engineered-in' defects: Case studies. *Engineering Failure Analysis, 12*(6), 1000–1026.

Goodman, I. (1914). *Mechanics applied to engineering*. London: Longmans, Green & Co.

Green, D., & Knowles, J. (1992). The treatment of residual stress in fracture assessment of pressure vessels. *ASME Journal of Pressure Vessel Technology, 233,* 237–247.

Hadrboletz, A., Weiss, B., & Stickler, R. (1994). Fatigue threshold of metallic materials–a review. In Carpinteri, A (Ed.), *Handbook offatigue crack propagation in metallic structures* (pp. 847–882). Elsevier.

Hensel, J., Nitschke-Pagel, T., Tchoffo-Ngoula, D., Beier, H.Th, Tchuindjang, D., & Zerbst, U. (2018). Welding residual stresses, fatigue crack propagation and fatigue strength. *Engineering Fracture Mechanic, 198,* 123–141.

Hobbacher, A. F. (2009). The new IIW recommendations for fatigue assessment of welded joints and components—a comprehensive code updated. *International Journal of Fatigue, 31,* 50–58.

Hobbacher, A. F. (2016). *Recommendations for fatigue design of welded joints and components (IIW collection)*. Heidelberg: Springer.

ISO 1099. (2006). *Metallic materials—Fatigue testing—Axial force-controlled method*. Geneva: International Organization for Standardization (ISO).

ISO 4287. (1997). *Geometric product specifications (GPS)—Surface texture: profile method—Terms, definitions and surface texture parameters*. Geneva: International Organization for Standardization (ISO).

ISO 5817. (2014). *Welding—Fusion welded joints in steel, nickel, titanium and their alloys (beam welding excluded)—Quality levels for imperfections*. Geneva: International Organisation for Standardization (ISO).

ISO 12106. (2017). *Metallic materials—Fatigue testing—Axial strain-controlled method*. Geneva: International Organization for Standardization (ISO).

ISO 12108. (2012). *Metallic materials—Fatigue testing—Fatigue crack growth method*. Geneva: International Organization for Standardization (ISO).

ISO 15653. (2010). *Metallic materials—Method for the determination of quasistatic fracture toughness of welds*. Geneva: International Organisation for Standardization (ISO).

Japanese Society of Materials Science (JSMS). (1998). Data handbook on fatigue strength of metallic materials. *Journal of the Society of Materials Science Japan, 1*(2.3).

Johnson, W. S. (2010). The history, logic and uses of the equivalent initial flaw size approach to total fatigue life prediction. *Procedia Engineering, 2,* 47–58.

Jonsson, B., Samuelsson, J., & Marquis, G. B. (2011). Development of weld quality criteria based on fatigue performance. *Welding World, 55,* 79–88.

Josi, G. (2010). Reliability-based management of fatigue failures (Ph.D. thesis). Canada: University of Alberta.

Junghans, E. (1998). Anwendung des Engineering Treatment Model für Mismatch (ETM-MM) auf Schweißverbindungen mit Berücksichtigung von Schweißnahtgeometrie und Werkstoffverfestigung (Ph.D. Thesis). TU Hamburg (in German).

Kim, Y.-J. (2002). Experimental J estimation equations for single-cracked bars in four-point bend: Homogeneous and bi-material specimens. *Engineering Fracture Mechanics, 69,* 793–811.

Kim, Y.-J., & Budden, P. J. (2001). Plastic η factor solutions of homogeneous and bi-material SE (T) specimens for toughness and creep crack growth testing. *Fatigue & Fracture of Engineering Materials & Structures, 24,* 751–760.

Kim, Y.-J., Kim, J.-S., Schwalbe, K.-H., & Kim, Y.-J. (2003). Numerical investigation on J-integral testing of heterogeneous fracture toughness testing specimens: Part I—weld metal cracks. *Fatigue & Fracture of Engineering Materials & Structures, 26,* 683–694.

Kiyak, Y., Madia, M., & Zerbst, U. (2016). Extended parametric equations for weld toe stress concentration factors and through-thickness stress distributions in butt-welded plates subjected to tensile and bending loading. *Welding World, 60,* 1247–1259.

Koçak, M., Seifert, K., Yao, S., & Lampe, H. (1990). Comparison of fatigue precracking methods for fracture toughness testing of weldments: Local compression and step-wise high R-ratio. In *Proceedings of International Conference on Welding-90*. Geesthacht, Germany : GKSS-Research Centre.

Koo, J.-M., Huh, Y., & Seok, C.-S. (2012). Plastic η factor considering strength mismatch and crack location in narrow gap weldments. *Nuclear Engineering and Design, 247,* 34–41.

Kucharczyk, P., Madia, M., Zerbst, U., Schork, B., Gerwin, P., & Münstermann, S. (2018). Fracture-mechanics based prediction of the fatigue strength of weldments Material aspects. *Engineering Fracture Mechanics, 198,* 79–102.

Kumar, V., German, M. D., & Shih, C. F. (1981). An engineering approach for elastic–plastic fracture analysis. EPRI-Report NP-1931, EPRI, Palo Alto.

Laham, S. A. (1999). SINTAP: Stress intensity factor and limit load handbook. British Energy Generation Ltd. (BEGL), Report EPD/GEN/REP/0316/98, Issue 1.

Lamba, H. S. (1975). The J-integral approach applied to cyclic loading. *Engineering Fracture Mechanics, 7,* 693–703.

Landgraf, R. W. (1970). *The resistance of metals to cyclic deformation*. ASTM STP 467 (pp. 3–36). West Conshohocken, PA: American Society for Testing and Materials (ASTM).

Lecsek, R. L., Yee, R., Lambert, S. B., & Burns, D. J. (1995). A probabilistic model for initiation and propagation of surface cracks in welded joints. *Fatigue & Fracture of Engineering Materials & Structures, 18,* 821–831.

Leggatt, R. H. (2008). Residual stresses in welded structures. *International Journal of Pressure Vessels and Piping, 85,* 144–151.

Lindgren, E., & Stenberg, T. (2011). Quality inspection and fatigue assessment of welded structures (Master Thesis). Stockholm: The School of Engineering Sciences (KTM).

Lopez, Z., & Fatemi, A. (2012). A method for predicting cyclic stress-strain curve from tensile properties. *Materials Science and Engineering: A, A566,* 540–550.

Maddox, S. J. (1970). Calculating the fatigue strength of a welded joint using fracture mechanics. *Metal Construction, 2,* 327–331.

Maddox, S. J. (1974). Assessing the significance of flaws in welds subject to fatigue. *Welding journal, 53*(9), 401–409.

Maddox, S. J. (1975). An analysis of fatigue cracks in fillet welded joints. *International Journal of Fracture, 11,* 221–243.

Madia, M. (2008): Effect of constraint on crack propagation in mechanical components (Ph.D. thesis). Politecnico di Milano.

Madia, M., Arafah, D., & Zerbst, U. (2014). Reference load solutions for plates with semi-elliptical surface cracks subjected to biaxial tensile loading. *International Journal of Pressure Vessels and Piping, 119,* 19–28.

Madia, M., Schork, B., Kaffenberger, M., & Bernhard, J. (2017). Multiple crack initiation and propagation in weldments under fatigue loading. *Procedia Structural Integrity, 7,* 423–430.

Madia, M., Thoffo Ngoula, D., Zerbst, U., & Beier, H.Th. (2017). Approximation of the crack driving force for cracks at notches under static and cyclic loading. *Procedia Structural Integrity, 5,* 875–882.

Madia, M., & Zerbst, U. (2016). Application of the cyclic R-curve method to notch fatigue analysis. *International Journal of Fatigue, 82,* 71–79.

Madia, M., Zerbst, U., Beier, H.Th, & Schork, B. (2018). The IBESS model—elements, realisation and validation. *Engineering Fracture Mechanics, 198,* 171–208.

Maierhofer, J., Kolitsch, S., Pippan, R., Gänser, H.-P., Madia, M., & Zerbst, U. (2018). The cyclic R-curve—determination, problems, limitations and application. *Engineering Fracture Mechanics, 198,* 45–64.

Marohnic, T., Basan, R., & Franulovic, M. (2017). Evaluation of methods for estimation of cyclic stress-strain parameters from monotonic properties of steels. *Metals, 7,* 17. https://doi.org/10: 3390/met7010017.

McClung, R. C. (1994). Finite element analysis of specimen geometry effects on fatigue crack closure. *Fatigue & Fracture of Engineering Materials & Structures, 17,* 861–872.

McClung R. C., Chell, G. G., Lee, Y.-D., Russell, D. A., & Orient, G. E. (1999). Development of a practical methodology for elastic-plastic and fully plastic fatigue crack growth. NASA Report NASA/CR-1999-209428.

McDowell, D. L. (1996). Basic issues in the mechanics of high cycle metal fatigue. *International Journal of Fracture, 80,* 103–145.

McEvily, A. J., Endo, M., & Murakami, Y. (2003). On the $\sqrt{\text{area}}$ relationship and the short fatigue crack threshold. *Fatigue & Fracture of Engineering Materials & Structures, 26,* 269–278.

Miedlar, P.C., Berens, A. P., Gunderson, A., & Gallagher, J. P. (2002). USAF damage tolerant design handbook: Guidelines for the analysis and design of damage tolerant aircraft structures. AFRL-VA-WP-TTR-2003-3002.

Miller, A. G. (1988). Review of limit loads of structures containing defects. *International Journal of Pressure Vessels and Piping, 32,* 197–327.

Miller, K. J. (1993). The two thresholds of fatigue behaviour. *Fatigue & Fracture of Engineering Materials & Structures, 16,* 931–939.

Minakawa, K., Newman, J. C., & McEvily, A. J. (1983). A critical study of the crack closure effect on near threshold fatigue crack growth. *Fatigue & Fracture of Engineering Materials & Structures, 6,* 359–365.

Moftakhar, A. A., & Glinka, G. (1992). Calculation of stress intensity factors by efficient integration of weight functions. *Engineering Fracture Mechanics, 43,* 749–756.

Monahan, C. C. (1995). Early fatigue crack growth at welds. In C. A. Brebbia & J. J. Connor (Eds.), *Topics in Engineering* (Vol. 26, pp. 1212–1216). Southhampton, U.K.: Computational Mechanics Publications.

Murakami, Y. (2002). *Metal fatigue. Effects of small defects and nonmetallic inclusions*. Oxford: Elsevier.

Murakami, Y. (2003). High and ultrahigh cycle fatigue. In R. O. Ritchie, Y. Murakami (Eds.), *Comprehensive structural integrity* (Vol 4, pp. 129–164). Cyclic loading and Fracture, Elsevier.

Murakami et al. (Eds.) (1987). *Stress intensity factor handbook.* Case 5.3 (Vol 1, 244ff). Oxford: Pergamon Press.

NASGRO. (2000). Fatigue crack growth computer program "NASGRO" Version 3, Houston, Texas: NASA.

Neuber, H. (1958). *Kerbspannungslehre: Theorie der Spannungskonzentration; Genaue Berechnung der Festigkeit.* Berlin: Springer. (in German).

Newman, J. C. (1981). *A crack closure model for predicting fatigue crack growth under aircraft spectrum loading*. ASTM STP 748 (pp. 53–84). Philadelphia: American Society for Testing and Materials (ASTM).

Newman, J. C. (1984). A crack opening stress equation for fatigue crack growth. *International Journal of Fracture, 24,* R131–R135.

Newman, J. C. (1992). FASTRAN II—a fatigue crack growth structural analysis program. NASA TM-104159.

Newman, J. C., Jr., Schneider, J., Daniel, A., & McKnight, D. (2005). Compression pre-cracking to generate near threshold fatigue-crack-growth rates in two aluminum alloys. *International Journal of Fracture, 27,* 1432–1440.

Newman, J. C., Ruschau, J. J., & Hill, M. R. (2010). Improved test method for very low fatigue-crack-growth-rate data. *Fatigue & Fracture of Engineering Materials & Structures, 34,* 270–279.

Nisitani, H., & Isida, M. (1982). Simple procedure for calculating K, of a notch with a crack of arbitrary size and its application to non-propagating fatigue crack. *Proceedings of Joint Conference on Experimental Mechanics Part I, 1,* 150–155.

Nitschke-Pagel, T. (1995). Eigenspannungen und Schwingfestigkeitsverhalten geschweißter Feinkornbaustähle (Ph.D thesis). TU Braunschweig (in German).

Nykänen, T., Li, X., Björk, T., & Marquis, G. (2005). A parametric fracture mechanics study of welded joints with toe cracks and lack of penetration. *Engineering Fracture Mechanics, 72,* 1580–1609.

Ohta, A., Suzuki, N., & Maeda, Y. (1997). Unique fatigue threshold and growth properties of welded joints in a tensile residual stress field. *International journal of fatigue, 19,* S303–S310.

Otegui, J. L., Burns, D. J., Kerr, H. W., & Mohaupt, U. H. (1991). Growth and coalescence of fatigue cracks at weld toes in steel. *International Journal of Pressure Vessels and Piping, 48,* 129–165.

Pang, H. L. J., & Gray, T. G. F. (1993). Fatigue analysis of surface cracks at fillet welded toes. *Fatigue & Fracture of Engineering Materials & Structures, 16,* 151–164.

Paredes, M., & Ruggieri, C. (2012). Further results in J and CTOD estimation procedures for SE (T) fracture specimens—Part II: Weld centreline cracks. *Engineering Fracture Mechanics, 89,* 24–39.

Peterson, R. E. (1959). Notch sensitivity. In G. Sines & J. L. Waisman (Eds.), *Metal fatigue* (pp. 293–306). New York: MacGraw Hill.

Pippan, R., & Riemelmoser, F. O. (2003). Modelling of fatigue growth: Dislocation models. In R. O. Ritchie, Y. Murakami (Eds.), *Comprehensive structural integrity* (Vol 4, pp. 191–207). Cyclic loading and Fracture, Elsevier.

Plekhov, O., Paggi, M., Naimark, O., & Carpinteri, A. (2011). A dimensional analysis interpretation to grain size and loading frequency dependencies of the Paris and Wöhler curves. *International Journal of Fatigue, 33,* 477–483.

Pokluda, J., Pippan, R., Vojtek, T., & Hohenwarter, A. (2014). Near-threshold behaviour of shear-mode fatigue cracks in metallic materials. *Fatigue & Fracture of Engineering Materials & Structures, 37,* 232–254.

Polak, J. (2003). Cyclic deformation, crack initiation, and low-cycle fatigue. In R.O. Ritchie, Y. Murakami (Eds.), *Comprehensive structural integrity* (Vol 4, pp. 1–39). Cyclic loading and Fracture, Elsevier.

Pusch, G., & Hübner, P. (1998). Bruchverhalten des Stahls StE960 und seiner Schweißverbindung bei statischer und zyklischer Belastung. Technische Forschung Stahl, Bericht EUR 17880 DE (in German).

R5, Issue 3. (2014). *Assessment procedure for the high temperature response of structures.* Barnwood, Gloucester: EDF Energy.

R6, Revision 4. (2014). *Assessment of the integrity of structures containing defects.* Barnwood, Gloucester: EDF Energy.

Radaj, D., Sonsino, C. M., & Fricke, W. (2006). *Fatigue assessment of welded joints by local approaches.* Cambridge, UK: Woodhead Publ LTD.

Radaj, D., & Vormwald, M. (2007). Ermüdungsfestigkeit. Springer, Section 3.1 (in German).

Rennert, D., Kullig, E., Vormwald, M., Esderts, A., & Siegele, D. (2013). *Analytical strength assessment of components made of steel, cast iron and aluminium materials in mechanical engineering. FKM Guideline* (6th ed). Frankfurt/Main, Germany: Forschungskuratorium Maschinenbau (FKM).

Rice, J. R. (1968). A path-independent integral and the approximate analysis of strain concentration by notches and cracks. *ASME Journal of Applied Mechanics, 35,* 376–386.

Ritchie, R. O. (1999). Mechanisms of fatigue-crack propagation in ductile and brittle solids. *International Journal of Fracture, 100,* 55–83.

Ritter, W. (1994). Kenngrößen der Wöhlerlinien für Schweißverbindungen aus Stählen. Report Institut für Stahlbau und Werkstoffmechanik der TH Darmstadt Vol 53 (in German).

Roessle, M. L., & Fatemi, A. (2000). Strain-controlled fatigue properties of steels and some simple approximations. *International Journal of Fatigue, 22,* 495–511.

Romano, S., Manenti, D., Beretta, S., & Zerbst, U. (2016). Semi-probabilistic method for residual lifetime of aluminothermic welded rails with foot cracks. *Theoretical and Applied Fracture Mechanics, 85,* 398–411.

Saxena, A. (1998). *Nonlinear fracture mechanics for engineers*. Boca Raton, Flo: CRC Press.

Schijve, J. (2004). *Fatigue of structures and materials*. New York : Kluwer Academic Publ.

Schork, B., Kucharczyk, P., Madia, M., Zerbst, U., Hensel, J., Bernhard, J., et al. (2018). The effect of the local and global weld geometry as well as material defects on crack initiation and fatigue strength. *Engineering Fracture Mechanics, 198,* 103–122.

Schwalbe, K.-H., Heerens, J., Zerbst, U., Pisarski, H. & Koçak, M. (2002). EFAM GTP 02—the GKSS test procedure for determining the fracture behaviour of materials. GKSS Research Centre, Report GKSS 2002/24, Geesthacht, Germany.

Schwalbe, K.-H., Kim, Y.-J., Hao, S., Cornec, A. & Koçak, M. (1997). EFAM ETM-MM 96—the ETM method for assessing the significance of crack-like defects in joints with mechanical heterogeneity (strength mismatch). GKSS Research Centre, Report GKSS 97/E/9, Geesthacht, Germany.

Schwalbe, K.-H., & Zerbst, U. (2003). Crack driving force methods. In R. A. Ainsworth, K.-H. Schwalbe (Eds.), *Comprehensive structural integrity* (Vol 7, pp. 136–176). Practical Failure Assessment Methods, Elsevier.

Schwalbe, K.-H., Zerbst, U., Kim, Y.-J., Brocks, W., Cornec, A., Heerens, J. et al. (1998). EFAM ETM 97—the ETM method for assessing the significance of crack-like defects in engineering structures, comprising the versions ETM 97/1 and ETM 97/2. GKSS Research Centre, Report GKSS 98/E/6, Geesthacht, Germany.

Sedlacek, G., Felmann, M., Kühn, B., Tschickardt, D., Höhler, S., Müller, C., et al. (2008). Commentary and worked examples to EN 1993-1-10 "Material toughness and through thickness properties" and other toughness oriented rules in EN 1993. JRC Scientific and Technical Reports, JRC 47278, Luxembourg.

Shih, C. F., & Hutchinson, J. W. (1976). Fully plastic solutions and large-scale yielding estimates for plane stress crack problems. *ASME-Journal of Engineering Materials and Technology, 9,* 289–295.

Shiratori M., Miyoshi T., & Tanikawa K. (1987). Analysis of Stress Intensity Factors for surface cracks subjected to arbitrarily distributed surface stresses. In Y. Murakami (Ed), *Stress Intensity Factors Handbook, chapter "Weight function for a semi-elliptical surface crack in a plate under basic mode of stress distribution*" (pp. 725–727). Oxford: Pergamon Press.

SINTAP—Structural Integrity Assessment Procedures for European Industry. (1999). *Project BE95-1426 Final Procedure*. Rotherham: British Steel Report.

Suresh, S. (2003). *Fatigue of materials* (2nd ed). Cambridge: Cambridge University Press.

Tabernig, B., & Pippan, R. (2002). Determination of the length dependence of the threshold for fatigue crack propagation. *Engineering Fracture Mechanics, 69,* 899–907.

Tan, J. T., & Chen, B. K. (2013). Coalescence and growth of two coplanar short cracks in AA7050-T7451 aluminium alloys. *Engineering Fracture Mechanics, 102,* 324–333.

Tanaka, K. (2003). Fatigue crack propagation. In: R. O. Ritchie, Y. Murakami (Eds.), *Comprehensive structural integrity* (Vol 4, pp. 95–127) Cyclic loading and Fracture, Elsevier.

Tanaka, K., & Akinawa, Y. (1988). Resistance-curve method for predicting propagation threshold of short fatigue cracks at notches. *Engineering Fracture Mechanics, 30,* 863–876.

Tanaka, K., & Akinawa Y. (1989). Propagation and non-propagation of small fatigue cracks. In K. Salama, K. Ravi-Chandar, D. M. R. Taplin, P. R. Rao, P.R. (Eds.), *Advances in fracture research. Proceedings of International Conference on Fracture (ICF) 7* (Vol 2, pp. 869–887). Houston (TX).

Tanaka, K., & Akinawa, Y. (2003). Modelling of fatigue crack growth: Mechanistic models. In R. O. Ritchie, Y. Murakami, Y. (Eds.), *Comprehensive structural integrity* (Vol 4, pp. 165–189). Cyclic loading and Fracture, Elsevier.

Tchoffo Ngoula, D., Beier, H Th, & Vormwald, M. (2018). Fatigue crack growth in cruciform welded joints: Influence of residual stresses and of the weld toe geometry. *Engineering Fracture Mechanics, 101,* 253–262.

Tchoffo Ngoula, D., Madia, M., Beier, H.Th, Vormwald, M., & Zerbst, U. (2018). Cyclic J-integral: Numerical and analytical investigations for surface cracks in weldments. *Engineering Fracture Mechanics, 198,* 24–44.

Tchuindjang, D., Fricke, W., & Vormwald, M. (2018). Numerical analysis of residual stresses and crack closure during cyclic loading of a longitudinal gusset. *Engineering Fracture Mechanics, 198,* 65–78.

To, S. H., Lambert, S. B., & Burns, D. J. (1993). A multiple crack model for fatigue in welded joints. *International Journal of Fatigue, 15,* 333–340.

Toyoda, M. (1989). Significance of procedure/evaluation of CTOD test of weldments International M. Document X-1192-89, Instituite of Welding (IIW).

Verreman, Y., & Nee, B. (1996). Early development of fatigue cracking at manual fillet welds. *Fatigue & Fracture of Engineering Materials & Structures, 19,* 669–681.

Verreman, Y., & Nie, B. (1991). Short crack growth and coalescence along the toe of a manual fillet weld. *Fatigue & Fracture of Engineering Materials & Structures, 14,* 337–349.

VOLVO Standard STD 181-0004. (2008). Volvo Group Weld Quality Standard.

Vormwald, M. (2014). Fatigue crack propagation under large cyclic plastic strain. *Procedia Materials Science, 3,* 301–306.

Vormwald, M., Heuler, P., & Seeger, T. (1992). *A fracture mechanics based model for cumulative damage assessment as part of fatigue life prediction.* ASTM STP 1122 (pp. 28–43). Philadelphia: American Society for Testing and Materials (ASTM).

Vormwald, M., & Seeger, T. (1991). The consequence of short crack closure on fatigue crack growth under variable amplitude loading. *Fatigue & Fracture of Engineering Materials & Structures, 14,* 205–225.

Walbridge, S. (2005). A probabilistic study of fatigue in post-weld treated tubular bridge structures (Ph.D. thesis). Canada: University of Alberta.

Wallin, K. (1998). Probabilistisk säkerhetsvärdering PROPSE—Materialparametrar, Rapport VALC444, VTT Tillverkningsteknik, 22.

Wallin, K. (2002). Master curve analysis of the "Euro" fracture toughness dataset. *Engineering Fracture Mechanics, 69,* 451–481.

Wallin, K., Nevasmaa, P., Laukkanen, A., & Planman, T. (2004). Master curve analysis of inhomogeneous ferritic steels. *Engineering Fracture Mechanics, 71,* 2329–2346.

Wang, X., & Lambert, S. B. (1998). Weight functions and stress intensity factors for semi-elliptical cracks in T-plate welded joints. *Fatigue & Fracture of Engineering Materials & Structures, 21,* 99–117.

Webster, S., & Bannister, A. (2000). Structural integrity assessment procedure for Europe–of the SINTAP programme overview. *Engineering Fracture Mechanics*, *67*(6), 481-514.

Wu, X. R., & Carlsson, A. J. (1991). *Weight functions and stress intensity factor solutions.* Oxford U.K.: Pergamon Press.

Wu, W. F., & Ni, C. C. (2007). Statistical aspects of some fatigue crack growth data. *Engineering Fracture Mechanics, 74,* 2952–2963.

Zerbst, U. (2016). Analytische bruchmechanische Ermittlung der Schwingfestigkeit. Final Report of IBESS Subproject A3 (in German). https://opus4.kobv.de/opus4-bam/frontdoor/index/index/docId/35989.

Zerbst, U., Ainsworth, R. A., Beier, H.Th, Pisarski, H., Zhang, Z. L., Nikbin, K., et al. (2014). Review on the fracture and crack propagation in weldments—a fracture mechanics perspective. *Engineering Fracture Mechanics, 132,* 200–276.

Zerbst, U., Ainsworth, R. A., & Madia, M. (2012). Reference load versus limit load in engineering flaw assessment: A proposal for a hybrid analysis option. *Engineering Fracture Mechanics, 91,* 62–72.

Zerbst, U., Ainsworth, R. A., & Schwalbe, K.-H. (2000). Basic principles of analytical flaw assessment methods. *International Journal of Pressure Vessels and Piping, 77,* 855–867.

Zerbst, U., Kiyak, Y., Madia, M., Burgold, A., & Riedel, G. (2013). Reference loads for plates with semi-elliptical surface cracks subjected to tension and bending for application within R6 type flaw assessment. *Engineering Fracture Mechanics, 99,* 132–140.

Zerbst, U., & Madia, M. (2015). Fracture mechanics based assessment of the fatigue strength: Aapproach for the determination of the initial crack size. *Fatigue & Fracture of Engineering Materials & Structures, 38,* 1066–1075.

Zerbst, U., & Madia, M. (2018). Analytical flaw assessment. *Engineering Fracture Mechanics, 187,* 316–367.

Zerbst, U., Madia, M., & Beier, H.Th. (2014). A model for fracture mechanics based prediction of the fatigue strength: Further validation and limitations. *Engineering Fracture Mechanics, 130,* 65–74.

Zerbst, U., Madia, M., & Hellmann, D. (2012). An analytical fracture mechanics model for estimating of S-N curves of metallic alloys containing large second phase particles. *Engineering Fracture Mechanics, 82,* 115–134.

Zerbst, U., Madia, M., Klinger, C., & Bettge, D. (2018b): Defects as a root cause of fatigue failure of metallic components. Part I: Basic aspects; Part II: Types of defects—non-metallic inclusions; Part III: Types of defects—cavities, dents, corrosion pits, scratches. Submitted to Engng. Failure Analysis.

Zerbst, U., Madia, M., & Schork, B. (2016). Fracture mechanics-based determination of the fatigue strength of weldments. *Procedia Structural Integrity, 1,* 10–17.

Zerbst, U., Madia, M., & Vormwald, M. (2017). Fatigue strength and fracture mechanics. *Procedia Structural Integrity, 5,* 745–752.

Zerbst, U., Madia, M., & Vormwald, M. (2018c). Applying fracture mechanics to fatigue strength determination—some basic considerations. To be submitted to International Journal of Fatigue.

Zerbst, U., Madia, M., Vormwald, M., & Beier, H.Th. (2018). Fatigue strength and fracture mechanics—a general perspective. *Engineering Fracture Mechanics, 198,* 2–23.

Zerbst, U., Schödel, M., Webster, S., & Ainsworth, R. A. (2007). *Fitness-for-service fracture assessment of structures containing cracks. A workbook based on the European SINTAP/FITNET procedure*. Amsterdam: Elsevier.

Zerbst, U., Vormwald, M., Pippan, R., Gänser, H.-P., Sarrazin-Baudoux, C., & Madia, M. (2016). About the fatigue crack propagation threshold of metals as a design criterion—a review. *Engineering Fracture Mechanics, 153,* 190–243.

MIX
Papier aus verantwortungsvollen Quellen
Paper from responsible sources
FSC® C105338

If you have any concerns about our products,
you can contact us on
ProductSafety@springernature.com

In case Publisher is established outside the EU,
the EU authorized representative is:
Springer Nature Customer Service Center GmbH
Europaplatz 3, 69115 Heidelberg, Germany

Printed by Libri Plureos GmbH
in Hamburg, Germany